Understanding
Sun Tzu

on the

Art of War

The Oldest Military Treatise in the World

By Robert L. Cantrell

Center For Advantage
Arlington, VA
www.centerforadvantage.com

Center For Advantage
P.O. Box 42049
Arlington, VA 22204

Copyright © 2003 by Robert L. Cantrell
Published by Center For Advantage
www.centerforadvantage.com

Printed in the United States of America

First Printing 2003

Library of Congress Control Number: 2002093398
ISBN 0-9722914-0-7

Attention Corporations, Educational Institutions, and Professional Organizations: This book is available at quantity discounts for educational and gift purposes. For information please contact - Center For Advantage, P.O. Box 42049, Arlington, VA 22204; info@centerforadvantage.com.

Contents

Dedicated to the soldiers

Preface

2,500 years ago a Chinese warrior and philosopher named Sun Tzu became a grand master of strategy and captured the essence of his philosophies in a book called, by English speaking nations, *Sun Tzu on the Art of War*. To this day, military strategists around the world have used Sun Tzu's philosophies to win wars and have made *Sun Tzu on the Art of War* a staple of their military education.

Those seeking to understand strategy in business, law, and life have also turned to *Sun Tzu on the Art of War* for the wisdom therein. For at the heart of Sun Tzu's philosophies are strategies for effective and efficient conflict resolution useful to all who wish to gain advantages over their opposition.

To read *Sun Tzu on the Art of War* is not to immediately understand it, however. Readers often find the text difficult to work with and Sun Tzu's philosophies counterintuitive to their day-to-day reality. This causes a dilemma for those who wish to teach and study Sun Tzu's philosophies. To simplify Sun Tzu's philosophies risks losing the subtlety of thought necessary to master *Sun Tzu on the Art of War*. To keep the subtlety risks losing the most basic insights to the complexity of the text.

Capturing the subtlety of Sun Tzu's philosophies in an understandable form is where *Understanding Sun Tzu on the Art of War* comes in. In this book, the author uses his many years of practical experience with Sun Tzu's philosophies to clarify them without simplifying them. He does so by examining contemporary battlefields within the context of Sun Tzu's philosophies and by examining Sun Tzu's own Taoist philosophical sources, sources that have been heard in popular Western culture through voices such as Yoda's in the *Star Wars* movies. A lot goes on between the lines in *Sun Tzu on the Art of War,* and that gives the book its power.

Author's notes:

1. The second half of this book contains a complete edited version of the original Lionel Giles translation of *Sun Tzu on the Art of War* that Luzac and Co. published in London and Shanghai in 1910. This edited version was designed to make the text clearer for modern readers than the dated original. Giles's translation, regardless, is still considered one of the best English translations available. Those with a more scholarly interest in Sun Tzu may also wish to read a copy of the Lionel Giles translation in its original form for which a number of sources exist both electronically and in hard copy.

2. This book uses masculine pronouns throughout the text. Such references pertain to both genders as applicable to a situation.

3. Any references that involve the commitment of bodily harm are intended to illustrate *Sun Tzu on the Art of War* for its originally intended purpose, war. Such illustrations are meant for use literally only by those government-sanctioned professionals authorized to undertake such activities as Sun Tzu described them, for example, military soldiers in the conduct of their duties. Those in other professions should take such references figuratively.

4. Excerpts from the *Tao Te Ching* are edited versions of a translation written by James Legge and published by Humphrey Milford in London in 1891. This editing is designed to make the text clearer for present day readers than the dated original.

Notes on the cover:

1. The cover art shows a spearhead that was hand forged in the ways characteristic of those used to create weapons throughout Asia and Europe in the pre-industrial ages.

2. The Chinese characters used on the cover appear on the original Lionel Giles publication of *Sun Tzu on the Art of War* from 1910. They translate literally to "Sun Tzu on military ways."

Sun Tzu said:

> The art of war is of vital importance to the State. It is a matter of life and death, a road to safety or to ruin. Therefore, it is a subject that must be thoroughly studied.

Introduction

Picture the rapids of a great river. See its waters rush over and around giant boulders. Close your eyes and listen to its roar. Then feel its relentless power when it crashes over a precipice. Now picture that you remove a cup of water anywhere along this river and sense how that water loses its power and starts to dry up in the sun. Then empty the cup back into the river, and know that as a part of the whole river, that water wears rocks into sand and does not dry up. So a soldier and philosopher observing a river from its banks in this fashion might hypothesize that a great army kept whole can conquer nations and still stay whole, but an army divided or too small will face peril and death. Any review he might make of successful military campaigns in the past and in his present would confirm his hypothesis. Like a river on its journey to the sea, he could therefore conclude that the way of fighting involves fighting as a unified whole, an entire army acting as one, with one objective in mind, and with its own preservation as an army also kept in mind.

This philosopher records his idea in writing on the bamboo strips of his time. Later, readers of his work see that his idea rings true regardless of their profession. A powerful and universal principle comes to light called the principle of winning whole, meaning winning with your resources and your objective intact. It represents the first of six universal principles described by Sun Tzu that, when used together as one, present the most powerful strategic method yet recorded in any profession for winning conflicts.

In this book, you have an invitation to take a journey that will uncover the heart of Sun Tzu's strategic method and how to apply it. This journey could change how you approach the many challenges you will face throughout the rest of your life. You will read about a martial art of the mind, a way to outthink and outfight those who oppose your needs and desires. Ideally, you will also gain an appreciation of the many ways opponents may direct force against you so you can position yourself accordingly to preclude harm.

THREE STEPS BACK

On this journey, we will take three steps back from the present day to review Sun Tzu's original work and representations of Sun Tzu's own source material. This material includes:

1. The philosophies of *Sun Tzu on the Art of War* as they apply to war.
2. Taoist philosophies, described best by the Chinese philosopher Lao Tzu in the *Tao Te Ching*, that form the apparent root of Sun Tzu's own philosophies and a core of traditional Chinese philosophical thought.
3. The Ways of life, drawn from the natural world, that form the apparent root of the Taoist philosophies described by both Lao Tzu and Sun Tzu.

We will use this material to uncover the meaning behind and the practical application of all six Sun Tzu principles that include:

1. Winning Whole - How to succeed with your resources and your objective intact
2. Leading to Advantage - How to prepare and position your soldiers for victory
3. Deception - How to keep your intentions secret from opponents
4. Energy - How to apply force effectively and efficiently
5. Strengths and Weaknesses - How to find the best path to the goal
6. Initiative - How to take and keep the advantage in a conflict

Our journey will conclude by combining these six principles into a powerful and unified philosophy that affords the means to plan, act, and succeed with the effectiveness and efficiency of one of Sun Tzu's own students. In so doing, we will discuss how Sun Tzu's strategic method applies to any profession, with illustrations drawn from business, law, medicine, sports, and personal relations. Enter first, however, into the world of the soldier that Sun Tzu knew.

THE SOLDIER

The soldier is the army and the army is the soldier. He wins by outthinking and outfighting his opponent with the best tools at his disposal. A thousand years ago, horses and armor, swords, shields, and lances represented the best tools for fighting at anyone's disposal. With them, men clashed in battle, outthought and outfought each other as individuals and as armies, and decided the fates of many great cultures.

Fast forward to the present day and these tools of war have long passed into history, replaced by weapons faster, more precise, and infinitely more destructive. Still, history can surprise us with the resiliency of the old. America's first major battle of the 21st century featured high tech Special Forces soldiers riding horses into battle with their Afghan allies much like the days of old – albeit supported by pilots in the likes of F-16s and F-18s instead of archers.

Regardless of the time or the place, the warrior within a man stays the same. He learns strategies that follow the principles of the ages. And no one, East or West, has better captured the essence of strategy then did Sun Tzu – 2,500 years ago.

DISSEMINATION

2,500 years ago, Sun Tzu's recorded words on military strategy proved so insightful that the Chinese nobility preserved them intact, and they, and many others who have read them to this day, would put those ideas to work. Translations of Sun Tzu's text spread throughout Asia. The ideas became a staple of Japanese military philosophy as firm as China's own and no doubt influenced battle plans to include the surprise attack on Pearl Harbor in 1941. A reading of *Quotations From Chairman Mao Tse-Tung* also reveals many direct paraphrases from Sun Tzu that Chinese, North Korean, and Vietnamese forces used militarily against the West and each other throughout the second half of the twentieth century. No doubt all these nations will continue to use Sun Tzu's ideas to gain military, political, and economic advantages appropriate to their national interests in the present day.

The use of Sun Tzu's philosophies in the West has a shorter and sketchier history. Some military historians suggest that Napoleon applied Sun Tzu's philosophies in his military planning and even carried a copy of Sun Tzu's book with him on his campaigns. He certainly could have had in his possession a French translation available in his time, and his methods of maneuver show a marked similarity to those described by Sun Tzu.

Historians also debate where and when the first deliberate application of Sun Tzu's philosophies by an English speaking commander took place. It may have been on the battlefields of Arabia in 1917. Here, T. E. Lawrence, a British officer leading an Arab army in a revolt against the Turkish army, used methods of maneuver warfare for which there is little doubt that Sun Tzu at least would have approved. Like Sun Tzu, Lawrence's use of maneuver warfare has been studied by military students the world over, and his actions serve to illustrate principles of effective warfare Sun Tzu described.

T. E. LAWRENCE

At the height of World War I, the Turkish army occupied the port city of Akaba on the Sinai Peninsula. This occupation posed a threat to British control of the Suez Canal further to the west.

At first glance, Akaba seemed invulnerable to British attack. From the sea, the British navy faced huge guns the Turkish army had installed on the cliffs that overlooked the city. These guns could sink any ship that dared sail too close. Behind Akaba, the British Army faced an expansive desert, considered uncrossable because of its punishing heat and absence of water.

Upon further reflection on Akaba, Lawrence apparently decided that Turkish commanders had allowed the strength of their position to also become its weakness. Since the Turkish army expected no threat from the desert, all their large guns faced the sea and could not be turned around. So in the summer of 1917, T. E. Lawrence led a contingent of men across the uncrossable desert. This contingent of men formed the core of an Arab army that attacked and captured Akaba from the desert side instead of the sea. Lawrence's attack surprised and defeated the Turkish garrison at Akaba. The great guns that only faced the sea proved useless to the Turkish defense. In the aftermath, most of the Turkish garrison was captured instead of killed. Lawrence's Arab army suffered few casualties of its own. Akaba itself remained intact.

In 1910, seven years before Lawrence's attack on Akaba, Dr. Lionel Giles, a staff member of the Department of Oriental Printed Books and Manuscripts at the British Museum in London, introduced the English-speaking world to an effective translation of Sun Tzu's writings. Lionel Giles published his translation through Luzac and Co. in London and Shanghai under the title, *Sun Tzu on the Art of War*. Lawrence, an avid reader with a keen interest in military strategy, had a seven-year span in which to discover and study *Sun Tzu on the Art of War* before his own military endeavors began.

Though we can at best only hypothesize that Lawrence studied *Sun Tzu on the Art of War*, we do know his method of attack on Akaba exemplified Sun Tzu philosophies: Lawrence attacked his enemy's weaknesses and won the battle whole. If Lawrence did in fact study *Sun Tzu on the Art of War*, then he fell into a minority. Though available since 1910, the study of *Sun Tzu on the Art of War* did not gain prominence in English speaking militaries until Vietnamese military leaders, to include Gen. Vo Nguyan Giap, used principles characteristic of Sun Tzu to defeat American forces in Vietnam in the 1960s and 1970s. Gen. Giap, known to have studied T. E. Lawrence and his military methods, was also known to have studied Sun Tzu.

PRESENT DAY USE

The study of *Sun Tzu on the Art of War* became an American military education staple after the Vietnam War. As a case in point, the U.S. Marine Corps book of strategy, *Warfighting,* builds upon ideas about maneuver warfare taken directly from *Sun Tzu on the Art of War*. Recently, business and law schools worldwide added *Sun Tzu on the Art of War* to their course of study. Why? The use of a weapon of the mind – which *Sun Tzu on the Art of War* fundamentally is – does not require the physical manifestation of the sword.

To read Sun Tzu's words, however, is not to immediately understand them. That would be equivalent to picking up a sword for the first time and expecting to fight well with it. To understand Sun Tzu's words means to put them into practice within the context of the Taoist philosophies from which he wrote them. These Taoist philosophies formed the cornerstone of philosophical thought for Sun Tzu's audience. They fill in the detail between the lines of Sun Tzu's own words. Sun Tzu described a lot more in his book than meets the eye. We will therefore take the next step of our exploration into China where history tells us an enlightened philosopher named Lao Tzu wrote a book called the *Tao Te Ching*.

THE *TAO TE CHING* AS A REPRESENTATION OF SUN TZU'S PHILISOPHICAL SOURCE

Tao Te Ching means "the Way of life." The philosophies described within the *Tao Te Ching* provide the foundation of the Taoist religion and the Taoist philosophy prominent in China during Sun Tzu's time. These philosophies continue to have such worldwide appeal today that only the Christian *Bible* is printed in a wider variety of languages.

Like *Sun Tzu on the Art of War*, scholars estimate the *Tao Te Ching* to be about 2,500 years old. Though a short book of only 81 verses, the *Tao Te Ching* influences Eastern views of the world similarly to the way the *Bible* influences Western worldviews. Most Western people, for example, regardless of their personal religious affiliations, understand the meaning behind the biblical philosophy, "turn the other cheek." Within these four simple words, you can communicate an important idea about conflict resolution that scholars have written about for centuries.

Philosophies from the *Tao Te Ching* similarly imbed themselves into Eastern culture. "Water seeks its own level," for example, communicates the inherent and often permanent good and bad of a person's nature, an idea much larger than the apparent sum of its five words. The phrase "the journey of a thousand miles begins with a single step," commonly used in both the East and West, is also from the *Tao Te Ching*. It communicates the humble origins of even the greatest endeavors. By using such understood ideas as his foundation, Sun Tzu succeeded in packing a lot of information into a very small text, particularly for those who have a foundation in Taoist ideas.

RECONCILING THE *TAO TE CHING* WITH WAR

Like the *Bible*, the *Tao Te Ching* acknowledges that a state's leadership cannot always avoid war. Lao Tzu therefore believed a military triumph should be treated as a funeral simply because war proved necessary at all. Sun Tzu, like any soldier trying to reconcile his religion with war, would have grappled with Lao Tzu's words against war. Lao Tzu said:

> *Arms, however beautiful, are instruments of ill omen, hateful to all creatures.*
> *Those who know the way of life do not wish to employ them.*
> *The superior man prefers his higher nature, but in time of war, will call upon his lower nature.*
> *Weapons are an instrument of ill omen, and not the instruments of the superior man, until he has no choice but to employ them.*
> *Peace is what he prizes; victory through forces of arms is to him undesirable.*
> *To consider armed victory desirable would be to delight in killing men, and he who delights in killing men will not prevail on the world.*
> *To celebrate when man's higher nature comes forth is the prized position; when his lower nature comes forth is time for mourning.*
> *The commander's second has his place in man's higher nature; the commanding general has his place assigned to man's lower nature; his place assigned to him as if to a funeral.*
> *He who has killed multitudes of men should weep for them; and the victor in battle has his place accorded as in a funeral.*

So you also have from Sun Tzu, in his third chapter, a philosophy that can reconcile the way to fight a war with the Way of life. Sun Tzu said:

> *In the practice of the art of war, it is best to take the enemy's country whole and intact. To shatter and destroy his country is inferior to this way. So, too, it is better to capture an army intact than to destroy it, better to capture a regiment, a detachment or a company intact than to destroy them. Hence to fight and win in all your battles is not the foremost excellence; to break the enemy's resistance without fighting is the foremost excellence.*

Throughout *Sun Tzu on the Art of War*, Sun Tzu advocated winning without actual fighting. The pursuit of that ideal, though not always possible, allows a warrior to reconcile military service with the *Tao Te Ching*. It also served a very practical purpose for Sun Tzu and his followers. In a world that contained many warring states, a victor who depleted his own resources to destroy an enemy could expect a third state to take advantage of his weakened condition. To win wars yet stay strong was essential to survival in China 2,500 years ago. Winning wars and staying strong carries equal importance today. And so we will begin our effort to understand Sun Tzu by exploring the first of six core principles discussed in this book – the principle of winning whole.

Six Principles of Sun Tzu

1 - Winning Whole

To win whole means to win with your resources and your objective intact. Any other result means you have at least partly failed at your mission. When you fight, you fight for something of value, and should you destroy yourself or that something of value while fighting to obtain it, then you have lost your real purpose for fighting.

The list below serves as a guide for winning whole. This chapter and the chapters that follow will develop this list.

How to Win Whole

1. **Remove your enemy's hope for victory**
2. **Use all your advantages**
3. **Exploit your enemy's weaknesses**
4. **Attack along an unexpected line**

If you deny your enemy any hope for victory, you diminish his will to fight, and defeating him will prove easier than otherwise. If you use all your advantages, you leave nothing on the table that might prevent undue loss. By attacking your enemy's weaknesses, the energy you must spend to succeed decreases. When you attack along an unexpected line, you can apply force against your enemy where he has not prepared for you. All these measures enhance your probability of winning with your resources intact and your objective intact. The Cold War, fought by the United States and its allies against the Soviet Union and its allies, illustrates these points.

THE COLD WAR

From 1945 to 1990, the United States and its allies fought a Cold War against the Soviet Union and its allies. This war threatened to become the most devastating Hot War imaginable.

The United States and its allies won the Cold War with little bloodshed, particularly when compared to the preceding global conflict, World War II. The United States and its allies maintained a sufficiently strong conventional military force to deter Soviet aggression and supplemented that force with nuclear weapons. The victors indirectly won the Cold War economically. The Soviet Union collapsed on itself without a direct military attack by the United States or its allies. Economies under Soviet-style communism could not afford to pay for the enormous militaries arrayed against the West. Bottom line: the Cold War ended and the world stayed whole.

THE AMERICAN CIVIL WAR

In contrast, America's Civil War of 1861-1865, though it succeeded in holding the union of American states together, demonstrated a progression into all the war-fighting methods Sun Tzu said to avoid. Sun Tzu said:

> *Thus the highest form of generalship is
> to defeat the enemy's plans;*

In 1861, Abraham Lincoln, the newly elected President of the United States, could not find a way to peacefully defeat the secessionist plans of Jefferson Davis, the President of the Confederate States of America. When Confederate forces shelled Fort Sumter in Charleston Harbor, they sealed Lincoln's failure.

> *...the next best is to keep the enemy's
> forces divided;*

From 1861 through 1863, Lincoln attempted to divide the Confederacy by taking control of the Mississippi River. At the same time, he made no fewer than three attempts to march on the Confederate Capital in Richmond, Virginia. Lincoln failed in part because Confederate officers, like Gen. Robert E. Lee, outclassed their Union counterparts on the battlefield.

> *...the next best is to attack the enemy's
> army in the field;*

In 1864, Lincoln assigned Gen. Ulysses S. Grant, a general who had earlier succeeded in capturing the key control city on the Mississippi, Vicksburg, to target Lee's Army of Northern Virginia itself instead of Richmond. Grant sent Gen. William T. Sherman down through the west to continue with the plan to divide the Confederacy. To Gen. George G. Meade of the Army of the Potomac, who Grant sent to destroy Lee's army in Virginia, Grant commanded: "Wherever Lee goes, there you will go also."

> *...and the worst policy is to besiege*
> *walled cities.*

After a bloody series of battles that failed to destroy Lee's army, Grant laid siege to Lee's army in the fortified city of Petersburg from the summer of 1864 to the spring of 1865. His need to lay siege meant adding 10 months to an effort that was taking some 200,000 casualties to whittle Lee's army from 60,000 soldiers at the beginning of 1864 to fewer than 10,000 by its surrender in 1865. Sherman's advances in the meantime destroyed vital Confederate cities south of Richmond and Petersburg, plus much of the region's railways and infrastructure. Though the war reunited the nation, the south arguably required more than one hundred years to recover from the method.

THE WAY OF LIFE

The American Civil War ended with America intact but with the south in ruin. It became a precursor of total wars to come that would see the devastation of fighting escalate sharply. World War I became the devastation of armies, and World War II became a total devastation of great cities and populations as well. Militarily, the great nations of Europe, both the winners and the losers, became a shadow of their former selves on the world stage and so relinquished the title of world superpowers to the United States and the Soviet Union. Errors on both sides – Allied weakness that allowed Axis forces to conquer Europe and the Asian Pacific with little loss to themselves, and Axis miscalculations regarding their ability to hold their gains for the long term – resulted in more death and destruction by war than the world has yet seen. This final result was not in accord with the Way of life and therefore not in accord with the principles of Sun Tzu.

But what exactly is this "Way of life?" Men have pondered this question for years. George Lucas borrowed the idea of the Way for a Western audience when he created the philosophies of Yoda in the *Star Wars* movies. On the Way, which Lucas called "the Force," Yoda said:

> *Size does not matter. Look at me. Judge*
> *me by my size do you? And where you*
> *should not. For my ally is the Force,*

and a powerful ally it is. Life creates it,
makes it grow. Its energy surrounds us
and binds us....

The Way, as attributed to Taoist philosophy, does not allow its masters to move stones with their mind like Lucas's Jedi Warriors, but it does purport to describe an energy that binds all things. Life, which energy sustains and destroys, creates conflict. The Way of conflict in the natural world sustains growth in the whole system. Predator and prey live in balance with each other to the benefit of the whole system. Winning whole also ascribes to the Way of life because you succeed at resolving conflicts productively and without destroying yourself and the objective. Fighting in accord with the Way of life, when you cannot avoid fighting, allows you to win whole with your army and your objective intact because you take measures, which we will discuss, that ensure this result. On the Way, in the *Tao Te Ching*, Lao Tzu said:

Before Heaven and Earth, there was
something undefined yet complete,
formless, alone, constant, everywhere
and untiring, the mother of all things.
I know not its name so I name it "the
Way of life." I should prevail to call it
great, for it is in constant flow, becoming
remote yet returning in a circle.
Therefore the Way is great; Heaven is
great; Earth is great; and a wise man is
also great.
In the universe, these are the four great
things.
Man takes his law from the Earth; the
Earth takes its law from Heaven; Heaven
takes its law from the Way; the law of
the Way being what it is.

A proper assessment of a situation is a critical part of Sun Tzu's overall philosophy that ensures an army engages in battles it can win. Engaging in battles you cannot win is a waste of time and resources and not in accord with the Way of life. In the first chapter of *Sun Tzu on the Art of War*, therefore, Sun Tzu described five constant factors a military leader must assess to afford a view of the whole situation before a battle. These constant factors correspond to the four great things of the universe we just read from the *Tao Te Ching*. The first constant factor corresponds to the Way of life. The second corresponds to heaven. The third corresponds to earth. The fourth and fifth correspond to man. Sun Tzu said:

*Five constant factors govern the art of
war, to be assessed when determining the
conditions in the field. These constant
factors are: 1) The Way 2) Heaven 3)
Earth 4) The Commander 5) Method
and discipline.*

The Way, which Lionel Giles translated into English as "Moral Law" for Western readers, proves the most important assessment of Sun Tzu's five constant factors prior to a military conflict and also the assessment easiest to miscalculate or ignore. The Way – Moral Law – determines whether one army can defeat another army without completely destroying it. Sun Tzu said:

*Moral law causes people to be in
complete accord with their ruler and to
follow him regardless of any danger to
their lives.*

The people of England had the moral law during the Battle of Britain in World War II where Winston Churchill said:

*Let us therefore brace ourselves to our
duties, and so bear ourselves that if the
British Empire and Commonwealth last
for a thousand years, men will still say,
this was their finest hour.*

An army prepared to defend its beloved homeland from capture, or an army cornered and fighting for its life, when compared to an army fighting without due cause, will find its soldiers' commitment to a fight more in line with the Way of life. Such an army may prove impossible to defeat in battle short of total annihilation.

An army imbued with the moral law is simply more willing to fight on and fight to the death than one without the moral law. To defeat an enemy with high morale while staying whole means to kill him, capture him, or break his morale; before soldiers meet in combat. Leave him hope and you invite disaster upon yourself. Winston Churchill said of the challenge that faced his enemies:

*We shall fight on the beaches. We shall
fight on the landing grounds. We shall
fight in the fields and in the streets. We
shall fight in the hills. We shall never
surrender!*

MORALE

Morale has four conditions that determine whether you or an enemy will fight or will quit. A key objective when fighting is to keep your morale high and your enemy's morale low. Of course, your enemy will seek the same against you.

Four Conditions of Morale

1. **The soldier fights – Because he has hope**
2. **The soldier quits – Because he can quit instead of fighting and still live**
3. **The soldier fights – Because he has no choice but to fight or die**
4. **The soldier quits – Because he loses his will to fight**

To foster high morale, keep in mind that soldiers fight because they have hope or because they have no choice but to fight or die. Also note that both of these reasons to fight can transcend a soldier's own self to include the hope or desperation of other people or an ideal, or both. This ability to transcend self allows a soldier to fight to his own death to save others when he might otherwise have surrendered and saved himself.

On the other side of morale, soldiers quit because quitting seems a better option than fighting, even if quitting means certain death. This too can transcend a soldier's own self if fighting for other people or an ideal loses its purpose.

To win whole, you find the means to keep your morale high and the morale of your charges high while you destroy your enemy's morale and make it easy for him to quit or impossible to escape destruction. You and any other military leader must also desire that your soldiers fight for the hope of some betterment to themselves or those for whom they care, but Sun Tzu knew a commander might have to leverage the hand of peril to succeed at the critical moment of battle. For this, he advocated that a commander leave no route for his own soldiers to return, and, therefore, no easy way for his own troops to quit. Sun Tzu said:

> *On the day they are ordered out to battle,*
> *your soldiers may weep, those sitting up*
> *crying into their garments, and those*
> *lying down letting the tears run down*
> *their cheeks. But let them once be in a*

situation where they cannot take refuge,
and they will display the courage of a
Chu or a Kuei.

At the heart of this statement may lay the resolve shown by Allied soldiers that landed on the Normandy beaches on D-Day on June 6, 1944. To get off the beach, they had no where else to go but forward. Allied commanders surely knew the implications of that fact before the attack began.

On the other side of morale again, history presents many cases where opposing commanders have made it difficult for their opposition to quit and have not destroyed their opposition's will to fight to compensate for this. Their policies served to put enemy soldiers in a place where they could not take refuge with the result that they fought harder than they might have otherwise. Russian soldiers, for example, would not surrender to the German SS during World War II. Surrender meant their certain execution, while fighting, no matter how desperate, at least afforded them a chance to live. In turn, the Russians would not take German SS soldiers as prisoners. Both sides suffered high casualties and high levels of brutality as a result. At a time when many Russians initially viewed German invaders as liberators over an oppressive Stalinist regime, military theorists must consider how World War II might have turned out if German forces had acted as liberators instead of proving an even greater evil.

In contrast, making it easy for an enemy to quit has helped America reduce casualties in most of its battles, particularly when by other means America had destroyed its opponents' will to fight. Take note of the mass Iraqi surrender during the Gulf War in 1991, after those Iraqi units had been subjected to the relentless American bombing campaign on their positions. Prisoners of the American military often received – and still receive – better treatment than what they received from their own armies. The underlying idea provides another way to win whole that Sun Tzu acknowledged when he said:

When you surround an army, leave an
outlet free. Do not press a desperate foe
too hard.

A desperate foe will fight back when you would prefer that he surrender. The general idea carries over into other aspects of life as well. It is a reason, for example, to find a way to "save face" in Asia so that an opponent can accept a loss without further conflict or disgrace.

MORALE TRAP

Of course high morale alone does not win wars when it leads to overconfidence. You may remember this excerpt from a certain famous poem by Alfred Lord Tennyson.

> *Half a league, half a league,*
> *Half a league onward,*
> *All in the valley of Death*
> *Rode the six hundred.*
> *"Forward, the Light Brigade!*
> *Charge for the guns!" he said:*
> *Into the valley of Death*
> *Rode the six hundred.*

You could argue that the soldiers of the British Light Brigade, about whom Alfred Lord Tennyson wrote these lines in 1870, had plenty of morale to stand on when the Cossacks and the Russians cut them down. Over 600 cavalry soldiers under Maj. Gen. the Earl of Cardigan followed orders to charge heavily defended Russian artillery positions over a mile distant from their starting point. The bravery displayed by these soldiers inspired Tennyson whose poem made them famous the world over. But despite the fame, Tennyson's poem describes a military fiasco. Perhaps 200 of those who set out on the charge returned alive with nothing to show for their effort. The Light Brigade never had a chance to succeed.

High morale, perhaps, and almost certainly overconfidence on the part of the commander, affected the two hundred and sixty troops belonging to U.S. Lt. Col. George Armstrong Custer when they fought Sioux, Cheyenne, and Arapaho warriors to the death at Little Bighorn on June 25, 1876. A commander must bear in mind certain other practical issues aside from morale before engaging in battle.

Consider Custer's defeat in context with Sun Tzu's other four assessments. Though heaven likely did not favor either Custer's cavalry or the Native American warriors in June of 1876, the conditions of earth would normally have favored the mobility of Custer's cavalry since the Native American warriors had also to protect their families and Custer's cavalrymen were not so encumbered. However, Custer's arrival in Montana, ahead of a supporting brigade-sized force, and Custer's subsequent decision to engage Native American warriors before the rest of that force arrived, gave Native American warriors an advantage of earth at a decisive point in time.

We do not know whether Custer underrated the Sioux Chief Sitting Bull's ability as a commander or underrated the method and discipline of his warriors. We do have evidence Custer discounted reports from his scouts that indicated Sitting Bull had assembled a large warrior force, a total that would number more than 2,000. This evidence is supported further by Custer's decision to proceed deeper into the valley which led to the hill where he fought his last stand despite losing contact with two flanking elements of his force.

In the end, the results of the battle show Custer had miscalculated the threat facing his men. Only one of Custer's troop, a horse named Comanche, survived when the Battle of Little Bighorn ended – though thirty men from one of his flanking detachments, not directly involved in Custer's fight, did return from the field. The final outcome for Custer proved the polar opposite of Sun Tzu's principle to win whole.

To avoid the morale trap, therefore, you must ensure you are not operating in accord with one of three conditions shown below. Think about this closely. The propensity to deceive ourselves on these matters is a major and unfortunate part of the human condition.

The Morale Trap

1. **You believe you have hope when there is no hope**
2. **You believe you fight to the death when you have other acceptable options**
3. **You allow pride to override good judgment**

If you believe you have hope when you do not, you may take ill adventures where you should not. If you believe you fight or die when you do not really face such dire conditions, you may needlessly continue a fight for no gain. If you allow pride to override your good judgment, you may fight even when you know you should not, regardless of the consequences.

With the above in mind, Custer's failure at the Battle of Little Bighorn prompts us to examine the second of the six Sun Tzu principles we will explore, leading to advantage. If you can avoid the morale trap, leading to advantage becomes a key part of winning whole.

2 - Leading to Advantage

With assessments completed in accord with Sun Tzu's five constant factors, a commander should understand the inherent advantages and disadvantages held by opposing armies. He may determine that one army holds overwhelming inherent and situational advantages and must defeat the other. But contested situations rarely produce such absolute conditions. So Sun Tzu discussed how military leaders can devise winning strategies from fortuitous or created advantages. You can therefore view Sun Tzu's five constant factors as a sort of pre-game assessment for war, while his fortuitous and created advantages provide the variable factors an army can leverage to actually win.

For a comparative illustration, if a star player on an opposing sports team catches the flu on game day, then your team has received a fortuitous advantage. You can use this fortuitous advantage by running plays against the star player's weaker substitute. Also, if you send your opponent off in the wrong direction during a play with a deception, then you have created an advantage within that play. You can use this created advantage to take another step toward victory as well. On this matter in war Sun Tzu said:

While heeding to the merit of my counsel
– counsel being the assessment on Sun
Tzu's five constant factors – *avail*
yourself also to any helpful
circumstances that give you advantages
beyond the ordinary conditions.

When such favorable circumstances
arise, modify your plans accordingly.

COWPENS

At the Battle of Cowpens, on January 17, 1781, 800 American soldiers under Gen. Daniel Morgan faced 1,100 British soldiers under Col. Banastre Tarleton in a battle to control the Carolinas. In this battle, Tarleton's soldiers, armed with muskets and bayonets, had an inherent advantage in close combat. Only half of Morgan's soldiers, and therefore one third of Tarleton's number, were similarly equipped with muskets and bayonets on that day. The rest of Morgan's soldiers were militiamen armed only with their personal hunting rifles. Though more accurate than muskets, hunting rifles took too long to reload for close combat and did not have bayonets.

To compensate for his apparent weakness, Morgan devised a plan to create an advantage for his men under the circumstances that would make the best use of the militia and their rifles. When Tarleton attacked, Morgan's militia soldiers, positioned in front of his Continental regulars, fired at least one aimed shot each at British officers and soldiers, as if they were hunting. When the British closed on them, Morgan's militia retreated behind the musket and bayonet armed line of Continental soldiers to reload. Then – something new for a militia – they re-entered the battle.

In accord with Morgan's plan, Tarleton saw the militia's retreat not as a military maneuver to create an American advantage, but as an opportunity to win a decisive British victory in the Carolinas. American militia had always retreated from British assaults in the past and left the battlefield. By Morgan's good fortune, a few of his Continentals also retreated in the confusion of the battle, and this fortuitous advantage actually completed Morgan's deception. Tarleton dispatched cavalry and ordered his infantry to charge.

From out of the smoke, the charging British soldiers encountered an organizing line of American militia and Continental troops. Instead of charging to a British victory, Tarleton's soldiers charged into the fields of fire of American muskets and rifles. After the ensuing American volleys, the battle became a decisive American victory. Morgan also won with his army intact.

Now ultimately, a leader's men, such as Morgan's, must fight a battle themselves. A good military leader finds advantages where his men's efforts deliver victory. He also inspires his men to use those advantages. On leadership, Lao Tzu said:

> At one time the people barely knew a leader's presence. In the next they loved and praised him.
> In the next they feared him. In the next they despised him.
> Thus it was that when a leader knew not the Way, the people's faith in him diminished.
> But of him with some humility; showing calm reserve and speaking little, a leader's work is done and his undertakings successful, while the people say we succeeded ourselves.

The victory Morgan's men could say they earned themselves directly reflected Morgan's own ability as their leader.

THREE ADVANTAGES

So in summary, when leading to advantage, you consider three types of advantages in your planning. These advantages follow below:

The Three Advantages

1. **Those inherent**
2. **Those fortuitous**
3. **Those created**

Inherent advantages describe what you have at the outset of an endeavor and correspond to the five constant factors described by Sun Tzu. Fortuitous advantages describe conditions that enhance your inherent advantages at a given place and time. Created advantages describe anything you do to enhance your inherent advantages and enhance your probability of benefiting from good fortune. The combination of these three advantages represents your total opportunity. Leading to advantage involves assisting individuals and teams to enhance all these aspects of advantage while in pursuit of your common goal. In consideration of his actions, and while directing others, a leader considers how to leverage the following list:

How to Lead to Advantage

1. **Assess inherent advantages in accord with the situation**
2. **Act in accord with the total opportunity**
3. **Enhance associated variables for either or both of the above**

A proper assessment of inherent advantages and a continual awareness of the total opportunity afford the best chance to act well when a critical moment to act arrives. Prior to the arrival of a critical moment, you have the opportunity to enhance the advantages you have. Training enhances inherent advantages. Increasing the quality, quantity, or both the quality and quantity of opportunities to succeed, perhaps by changing location or changing operational methods, enhances the total opportunity. However, leading to advantage does not stop here. A leader can do even more to make improvements on advantages deliver. For that, we return to Sun Tzu and his philosophies on war.

INVULNERABILITY

The ultimate advantage in war is invulnerability. The leader of an invulnerable army can choose when, where, and whether to fight a battle. Sun Tzu said:

> *The good fighters of old first made themselves invulnerable to defeat, and then waited for an opportunity to defeat their enemy.*

So what exactly did Sun Tzu mean by invulnerability? Does invulnerability mean the enemy cannot reach you? Perhaps it does for a time. But when given enough time, your enemy will find a way, a fact punctuated by the attacks on the United States that occurred on September 11, 2001. Does invulnerability mean bringing overwhelming force to bear against an enemy? Not necessarily. Many an army has defeated a superior force once engaged on the battlefield, a fact punctuated by the 1968 Tet Offensive that the United States won on the Vietnamese battlefields and lost with the population at home. Invulnerability means foresight. For example, think about how you might defend your house from flooding. No matter how many sandbags you have and how well you place them, nothing makes your house less vulnerable to flooding than having the foresight to build it on a hill.

Likewise, nothing makes you less vulnerable to an enemy than not being a target. You either make yourself too valuable to an enemy to attack, or deny an enemy the opportunity, the reason, or the ability to attack you. Then, if you have the opportunity, the wherewithal, and a reason to do so of your own, you can attack your enemy, or lead him to attack you, on ground and under circumstances of your own choosing.

PREVENTION AND INVULNERABILITY

Invulnerability begins with the prevention of hostilities. Prevention is the preferred method to attain invulnerability since it precludes any need to fight. Prevention does not mean appeasement since appeasement tends only to delay war. Prevention instead means removing the conditions that could lead to war. We find this aspect of invulnerability described in the *Tao Te Ching*. Lao Tzu said:

> *Before it moves, hold it; before it spoils,*
> *fix it; the brittle is easily broken; the*
> *small still easy to disperse.*
> *Take action before a bad thing happens;*
> *secure order before disorder begins.*

Lao Tzu's philosophy basically describes the common wisdom that:

> *An ounce of prevention is worth a pound*
> *of cure.*

Let no danger evolve that you can foresee, and you, by default, become less vulnerable. An enemy made into a friend reduces vulnerability. An enemy disarmed also reduces vulnerability, as long as that disarmament does not give rise to a greater danger elsewhere.

NEUTRALIZATION AND INVULNERABILITY

Though preventing any need for hostilities by default reduces vulnerability, we are not always blessed with the perfect foresight and the cooperativeness of neighbors such prevention requires. So the need to secure invulnerability through the neutralization of an enemy may arise instead.

On matters of neutralization, which implies the use or the threat to use force, Sun Tzu emphasized the importance of defeating an enemy before engaging him in battle. On matters of fighting, Sun Tzu clearly favored using means other than actual combat to win battles when those means produced the end result for which combat would have taken place. This sentiment appears in the aforementioned verse, first noted in the introduction, when Sun Tzu said:

> *Hence to fight and win in all your battles*
> *is not the foremost excellence; to break*
> *the enemy's resistance without fighting*
> *is the foremost excellence.*

When holding to the above ideal fails, and actual combat must take place, continuing to ascribe to this verse without compromising the integrity of the mission still provides a base from which to plan and take action. Your goal is to neutralize an enemy's capacity to fight before you commit the main body of your force to the battlefield. Should a conflict escalate to the point where you must commit the main body of your force, you strive to effectively defeat your enemy before your main body engages in close-quarter combat by preparing the battlefield for your attack. Through overt and covert means, you make the enemy physically, psychologically, and logistically unable to resist the advance of your army. Your enemy has no chance to win. Sun Tzu said:

> *He wins his battles by making no*
> *mistakes. Making no mistakes brings*
> *certain victory, for it means conquering*
> *an enemy that is already defeated.*
> *Hence the skillful fighter takes a position*
> *that makes his defeat impossible, and*
> *does not miss the opportunity to defeat*
> *his enemy. Thus it is that in war the*
> *victorious strategist seeks battle after the*
> *victory has been won, whereas he who is*
> *destined to defeat first fights and*
> *afterwards looks for victory in the midst*
> *of the fight.*

You can hardly overemphasize the importance of this idea from Sun Tzu. The idea to attack an enemy only after he is already defeated is so important in winning conflicts that it also fits into the doctrine of every predator in the wild. Predators live in constant conflict with their prey since their prey will take whatever action possible to avoid being eaten. Predators cannot, in this conflict, afford to subject themselves to a high risk of injury from prey that fights back since the odds against their own long-term survival would catch up with them too fast. Top predators therefore prefer to use overwhelming force against their prey. In the presence of that overwhelming force, they take further measures to reduce what risk they still have. Even a fifteen-foot white shark will ambush a five-foot seal from below, bite it, and let it bleed to death before feeding lest the shark risk injury from the seal's jaws. The shark cannot avoid all risk. It would starve. But it certainly takes no more risk than necessary to obtain its meal. By the shark's use of the method of ambush, the seal never has a chance to fight back.

Like a predator, a military leader must have eminent powers to detect and track his enemy's whereabouts, intentions, and moments of maximum vulnerability. Knowing an enemy's whereabouts and intentions in turn is also a key to fighting from an invulnerable position. Bottom line: if you know where an enemy is and he does not know where you are, and if you can destroy him and he cannot destroy you, then you can follow Sun Tzu's guidelines on neutralization and invulnerability. Foresight provides the means to make these guidelines real.

FORESIGHT AND INVULNERABILITY

Invulnerability, when you cannot prevent a conflict, means you at least have the ability to say "no" to the opposition without suffering undue consequences. It puts the power of decision into your hands.

Foresight – the means to see likely futures – provides for the ability to say "no" if you use it to prepare the means to say "no" in advance. Regarding foresight, Sun Tzu dedicated his last chapter to military espionage. Sun Tzu said:

> *What enables the wise sovereign and the good general to strike and conquer and achieve things beyond the reach of ordinary men is foreknowledge.*

Sun Tzu also emphasized the importance of obtaining intelligence directly from people – spies. To truly understand an enemy's intentions, Sun Tzu said, in no uncertain terms:

> *Now this foreknowledge – of the enemy – cannot be elicited from spirits; it cannot be obtained inductively from experience, nor can it be obtained by any deductive calculation. Knowledge of the enemy's disposition can only be obtained from other men.*

A spy serves his leadership best if his leadership knows how to make his state too useful in peace to attack in war. This usefulness in peace becomes a key to preventing war without undue appeasement. It basically involves one or more of the following five requirements:

Securing Invulnerability

1. **Know where a potential adversary is and what he is doing**
2. **Become more beneficial to a potential adversary as a friend than an enemy**
3. **Deny a reason for a potential adversary to attack you**
4. **Deny the opportunity for a potential adversary to attack you**
5. **Eliminate the adversary**

You can remember these five requirements by thinking about how you might live with a tiger in your house. First, you want to make sure you know where the tiger is and what the tiger is doing at all times. Second, if you can make the tiger your friend and become a part of its life, the tiger will not attack you and may even come to your defense if another entity attacks you. Third, if you cannot establish a friendship with the tiger, then you can perhaps deny it a reason to attack you by keeping it well fed and by not encroaching on its territory. Fourth, if two and three fail, then you might consider keeping the tiger or yourself inside a cage and the other outside the cage. If none of those conditions prove acceptable, and you have the means, you may have to kill the tiger.

Lao Tzu, in the *Tao Te Ching,* wrote about one country's safest place among its neighbors:

> *A great state is like a low-lying river; it becomes the center to which tend all small states under heaven; drawn to it like the female draws the male.*
> *The large state, if below the small state, will absorb the small state.*
> *The small state, if below the large state, wins over to the large state.*
> *The one case gains adherents, the other case procures favors.*
> *The great state wishes to unite and nourish men; a small state wishes to be received by and serve the other.*
> *Each gets what it desires when the great state learns to lower itself.*

This verse means that even a small entity can thrive among more powerful entities by establishing a place of mutual value the more powerful entity will not violate, and a powerful state, like the tiger that receives food from its physically weaker keeper, serves itself best by not also eating that keeper. Traditionally neutral Switzerland, for example, has had a banking system that allows even enemies to transact business with each other if necessary, and so maintains its neutrality peacefully. Hong Kong, under the protection of Great Britain, maintained its place as a center of commerce on the once hostile shores of China and later became a center point of business experience that has helped China to thrive as a market economy. Foreknowledge allows an entity to position itself in a safe place and, at a minimum, prepare for threats that may come in the best way possible.

PREPAREDNESS

If all nations shared the same rationale for mutual benefit that makes friendship possible, we might live in a peaceful world. However, not all enemies share our same rationale. Not all respond well to the same incentives we do. Naiveté can kill your finest. Therefore, a nation of peace must always have the ability to fight and kill when an enemy can only understand destruction. Sun Tzu said, therefore:

> *The art of war teaches us to not rely on*
> *the chance that the enemy will not come,*
> *but on our own preparations to receive*
> *him; not on the chance of an enemy not*
> *attacking, but rather on the fact that we*
> *have made our position unassailable.*

Preparation for conflict therefore requires you to consider the following:

Preparation Requirements

1. **Have the ability to act**
2. **Have the authority to act**
3. **Understand the consequences of acting**
4. **Understand the consequences of not acting**

To have the ability to act means you have the resources and the required skills to act. To have the authority to act means you can use that ability without undue restriction. Understanding the consequences of acting means you have assessed and can accept either success or failure associated with acting. To understand the consequences of not acting likewise means to accept either success or failure associated with not acting. Bottom line: we are all accountable for what we do or fail to do, and, in the absence of a decision, time tends to decide what we do for us. Also, those affected by the consequences of action or inaction tend to hold those with the ability to act more accountable for those consequences than others, particularly when that ability comes with the commensurate authority. This is true in war and otherwise.

MATERIAL ASPECTS OF FIGHTING

A leader must always consider the material aspects of preparedness to have the ability to act on opportunities and threats. Sun Tzu dedicated his second chapter to materials. In accordance with his day, he said:

> In the conduct of a war, where an army fields a thousand swift chariots, as many heavy chariots, and a hundred thousand mail-clad soldiers, with provisions enough to carry them a thousand li, the expenses at home and at the front, including entertainment of guests, small items such as glue and paint, and sums spent on chariots and armor, will reach the total of a thousand ounces of silver per day. Such is the cost of raising an army of 100,000 men.

No doubt the armies of Sun Tzu's time, like the armies of today, had their logistical manuals and past experiences to guide them on provisioning. Sun Tzu did not delve into the details beyond the above. Actually, Sun Tzu's discussion on logistics served as a warning. Sun Tzu said that no country has ever benefited from a prolonged war. He said:

> So in war, then, let a victory be your main objective, not the conduct of lengthy and costly campaigns.

Aside from the cost and the possible diminishment of your own morale, a prolonged war gives an enemy too much time to learn how to fight you. An enemy that learns how to fight you poses great danger. To defeat an enemy before that time, we move on to Sun Tzu's third principle, deception.

3 - Deception

An able enemy does not willingly lead his army to defeat. Hence, you have the role of deception in war. Deception affords your primary means to cause your enemy to make mistakes and expose himself to defeat.

In the opening chapter of *Sun Tzu on the Art of War*, Sun Tzu said:

> *All warfare is based on deception.*

In his chapter titled "Maneuver," Sun Tzu said:

> *In war, practice the art of deception and*
> *you will succeed.*

On both of these points, Sun Tzu referred to the importance of deception for obtaining victory. He supported his statements by describing how to create deceptions, generally through the use of stratagem and maneuver. Case in point, Sun Tzu discussed deception in terms of deviation in his following idea to deliberately entice an enemy off course while your own army takes an indirect route to the goal. Sun Tzu said:

> *Thus, to take an indirect route, after*
> *enticing the enemy out of the way, and*
> *though starting out after him, to contrive*

to reach the goal before him, you show
knowledge of the stratagem of deviation.

Now, though Sun Tzu said to deceive, and though he gave many examples of what to do to deceive, his text takes some reading between the lines to uncover how people are deceived. A place to start involves Sun Tzu's description of five dangerous faults of a general. Leveraging these faults in an opponent provides a means to devise deceptions. Sun Tzu said:

> *There are five dangerous faults that may*
> *affect a general:*
>
> *(1) Recklessness, which leads to destruction*
> *(2) Cowardice, which leads to capture*
> *(3) A hasty temper, which can be provoked*
>   *by insults*
> *(4) A delicacy of honor, which is sensitive*
>   *to shame*
> *(5) Too much compassion for his men, which*
>   *exposes him to worry and trouble.*
>
> *These are the five dangerous faults of a*
> *general, ruinous to the conduct of war.*
>
> *When an army is overthrown and its*
> *leader slain, the cause will surely be*
> *found among these five dangerous*
> *faults. Let them be a subject of*
> *meditation.*

A commander can use any of these five faults found in an opponent to devise a deception. For example, he can leverage an opponent's inherent recklessness to encourage him to charge into a trap. Gen. Daniel Morgan did this to Col. Banastre Tarleton when he created the illusion of retreat and so incited Tarleton to charge into his prepared line at the aforementioned Battle of Cowpens. But what do you do when an opponent does not have such obvious and exploitable faults? Sun Tzu advised, and we will explore further, about the importance also of leveraging that which an enemy needs or desires in order to deceive him. First, however, to find the heart of how people are deceived, we look at the *Tao Te Ching*.

HOW PEOPLE ARE DECEIVED

In the *Tao Te Ching*, Lao Tzu told how people are deceived. Lao Tzu said:

> *Men in the highest class, when they hear*
> *the Way, earnestly practice it.*
> *When of the middle class and they hear*
> *the Way, they sometimes keep to it and*
> *sometimes lose it.*
> *When in the lower class and they hear*
> *the Way, they laugh at it.*
> *If they did not laugh at it, it would not*
> *be the Way.*
> *To such who would laugh, the Way, when*
> *bright, seems dark; the smooth path*
> *appears rough; its highest virtue appears*
> *an empty valley; its greatest beauty*
> *offends the eyes; when at its most it*
> *seems insufficient; its greatest virtue*
> *seems poor.*

The susceptibility of people to deception depends on their willingness and ability to perceive the truth, the ultimate truth, according to Taoist beliefs, being called the Way. Lao Tzu ranked people by their ability to see the Way, and so the truth, when he separated men into the highest, middle, and lower classes.

Beyond any inherent susceptibility to deception, a person's willingness and ability to perceive the truth also revolves around that person's naiveté or experience. Those who have experienced a deception once tend to have a greater awareness of the possibility they might encounter that deception again.

The intensity of a need or desire will also influence a person's willingness to perceive the truth. The greater a person's need or desire, generally the less willing that person is to perceive an undesirable truth, regardless of that person's ability to do so.

So one knowledgeable about the Way of life is not easily deceived, and a commander knowledgeable about the ways of war is not easily fooled by another commander's deceptions. Like a lover who can tell when a lie crosses the lips of a partner, a commander familiar with the ways of war can sense a deception before his army falls into danger. The master of deception must always find new ways to deceive a wise enemy, and a soldier who is fooled by an old trick has only himself to blame for not knowing the truth that lay before him.

ALTERING THE APPARENT TRUTH

Altering the apparent truth that lies before an enemy is an art. Sun Tzu tells what your art should do to the enemy. Then, once you understand how people are deceived, you can read between the lines of Sun Tzu's statements and make the deceptive actions he described a reality. Case in point, Sun Tzu said on deception:

> *When able to attack, seem as if unable to attack; when using forces actively, seem inactive; when nearby, make the enemy believe you are far away; when far away, make the enemy believe you are nearby.*

To seem as if able or unable to attack, active or inactive, nearby or far away, you deceive in accord with the needs and desires of your enemy and also in accord with what he fears, which in turn plays directly into his willingness and ability to perceive the truth. Those who believe they have their need or desire in their grasp may lose their willingness and ability to perceive an alternative truth, and likewise, those affected by fear may magnify the actual danger you present in their own minds. These tendencies are a part of human nature and take some effort on the part of an individual to override, particularly when in a group of like thinking individuals, and particularly when that individual has one or more of a commander's five dangerous faults affected by the deception. In all counts, needs and desires forecast actions, and fear likewise causes action or inaction often associated with those needs or desires. When you know your enemy's needs and desires, when you know his fears, and when you further know your enemy's dangerous faults, you can leverage all those factors to make your deception complete. You can encourage your enemy to pursue gain or avoid the realization of his fear when in fact the action he pursues at best does nothing for him, and at worse, causes the opposite effect of what he intended.

HOW TO DECEIVE

Therefore, to deceive an enemy, do one or a combination of the actions from the following list:

How to Deceive

1. **Show your enemy what he wants to see**
2. **Show your enemy what he expects to see**
3. **Have your enemy see nothing**

If you show your enemy what he wants or expects to see, then, to the limit of his ability and willingness to perceive otherwise, your enemy will believe your deception. If you show him nothing, then he may not even know about your deception until it has already worked. Note, however, that showing an enemy nothing is often impossible. So the real art here involves not minding if an enemy sees something because, even if he does, he cannot perceive the truth before his eyes. If you reveal and conceal selectively, you can create an elaborate strategy of overall deception.

SOVIET DOCTRINE

Throughout the Cold War, the Soviet Army ran massive training exercises on the border of West Germany in plain sight of NATO observers. NATO military planners fully expected that, should the Soviets launch an invasion of Western Europe, they would launch it from one of those exercises. In no other way could the Soviets move the required divisions to the West German border without arousing undue suspicion. Following the Soviet doctrine from which they had trained, the Iraqi Army successfully enacted this deception in 1990, when a scheduled training exercise on the border of Kuwait became the surprise invasion of Kuwait.

PICKETT'S CHARGE

Two days of intense fighting on the right and left sides of the Union line at Gettysburg on July 1 and 2, 1863, left Union Gen. Winfield S. Hancock, in command at the center under Gen. George G. Meade, convinced that Gen. Robert E. Lee would attack that center the next day. On July 3, 1863, Confederate shelling at the center, with the apparent intent to destroy Union artillery batteries placed there, confirmed Hancock's supposition.

A short time after the Union and Confederate artillery dual commenced, Hancock ordered most of his cannons to cease firing. Hancock wanted the Confederates to believe they had destroyed these cannons, when in fact, Hancock used the lull to draw his guns into better positions along the line. There, he had his artillerymen prepare to fire short range but highly lethal canister rounds – lead shot packed in sawdust that scattered like a shotgun blast – into the expected Confederate infantry charge.

After the two previous days of fighting, the Confederate artillery units had depleted much of their ammunition. Lee needed a quick and decisive success before he had to turn his army back south. So when Hancock's cannons stopped firing, the Confederate commanders executing Lee's plan on the line saw what they wanted and expected to see as a result of their efforts, though Hancock's batteries had actually received little damage from their fire.

When a Confederate force of 15,000 men emerged from the woods and marched across the open field, Hancock ordered his artillerymen to open fire with all their cannons. In the engagement that ensued, Hancock's men effectively destroyed their Confederate attackers, to include Gen. George E. Pickett's division for which the attack received its name.

MYSTERY

Along with the basics on how to deceive appears a key to enacting deceptions well. This key is called mystery. It means that an enemy cannot know you even if you know him and therefore cannot know how or where to fight you. Lao Tzu, in the *Tao Te Ching*, afforded a precedent for mystery. Lao Tzu said:

> *I have heard that he who is skillful in managing the life entrusted to him for a time travels without shunning the rhinoceros or tiger, and enters a host without avoiding armor plate or sharp weapons.*
> *The rhinoceros finds no place in him to thrust his horn, nor the tiger a place to fix its claws, nor weapon a place to thrust its point.*
> *And for what reason?*
> *Because there is in him no place to die.*

Your enemy cannot kill what he cannot find. He cannot counter what he cannot understand. If your enemy cannot find you or understand your ways, you become a mystery to him. He cannot act well against you. It becomes easier to deceive him because he will develop his own ideas about you, likely biased by his own desires and fears, to fill in for what he cannot actually detect. This tendency, also a part of human nature, provides fertile ground to show him what you would like him to perceive. On mystery, in *Sun Tzu on the Art of War*, Sun Tzu said:

*All men can see the tactics whereby I
conquered my enemy, but none can see
the strategy I used to create the victory.*

Here's the subtlety between the lines in Sun Tzu's words. The actual mechanics of weapons use changes little with the circumstances. A soldier shoots a rifle, wields a sword, or fires artillery much the same way in all circumstances. The strategy that predisposes a soldier's enemy to defeat, however, has already succeeded before that soldier ever uses his weapon. Overtly or covertly, you have already destroyed your enemy's infrastructure and his morale, and yet the core of how you accomplished this remains a mystery. On destroying the enemy Sun Tzu said:

*Let your plans be dark and impenetrable
as night, and when you move, fall upon
your enemy like a thunderbolt.*

In war, mystery plays more on fear than desire. Therefore, the military leader creates the ultimate horror story for his enemy. As the best horror movie directors know, fear of the unknown is often worse than the unknown's realization. Waiting for the unknown sustains fear and destroys morale. Observing the effects of the unknown magnifies fear and the destruction of morale. Once realized, the unknown must devastate the enemy and achieve victory. Otherwise, an enemy can adapt to any known level of horror that does not kill him. To the degree your plans do not actually destroy your enemy's will or ability to fight, the realization of your plans may actually strengthen your enemy's resolve. If your enemy has any hope to believe his cause can win, even in his individual death, he may surprise you with the horror he can create in return.

EMPLOYING MYSTERY WITH DECEPTION

Therefore, to employ mystery with deception, do the following:

Deploying Mystery

1. **Physically and psychologically set the stage for battle**
2. **Maintain secrecy**
3. **Win before your opponent can adapt**

Again, think like the director of a horror movie and make your opponent your audience. Set the stage for the action you intend, both the physical surroundings and the psychology of the people involved. Think how Alfred Hitchcock built the famous shower scene in *Psycho*. He set the stage with the dramatic storyline, the nighttime storm, the isolated Bates Motel, the intense music, and an assortment of additional sights and sounds, all with the intent to maximize the shock the audience experiences. Next, to build apprehension and create a fear of the unknown, recall how Steven Spielberg sustained the fear in his audience throughout the movie *Jaws* by keeping the rogue shark present but hidden from view through most of the movie. You can imagine that even the suspected presence of an invisible stealth bomber at night attacks the psychology of an enemy in similar, if not more intense, ways when he must work or fight in a potential target zone. Finally, when you do reveal yourself, make sure you kill your enemy – to be taken figuratively outside the military – before he can adapt and fight you. Failure to do this last step, of course, is the fatal flaw of all our Hollywood villains. They fail to kill our protagonists. When our protagonists finally figure out how to kill them, our villains pay dearly as a result.

Bottom line: building mystery keeps your enemy guessing, and if you use mystery to enhance your deceptions, keeps your enemy guessing wrong. Like so much that is important in a conflict, nature shows us a precedent.

THE WOLF

The advantage inherent with mystery explains why the wolf goes for the jugular vein when it commits to attack a buffalo, yet keeps its distance until it commits to that attack and leaves the buffalo unsure where, when, and if it will actually attack. It also provides one explanation for why the wolf attacks in a pack, coordinating its attack with other wolves as one combined force. The buffalo, with its attention on one wolf, cannot act against other wolves when they attack unseen from different angles. Many wolves become as one to afford the best advantage to each individual, an idea Rudyard Kipling recorded in *The Law of the Jungle* when he wrote:

> *For the strength of the pack is the wolf,*
> *and the strength of the wolf is the pack.*

Should the wolf choose any other method of attack than the above, the wolf would expose itself to the defending buffalo's horns for too long. Worse, a surviving buffalo that learns ways to defend itself may serve to show other buffalos how to do the same at the expense of all wolves in the future. Sun Tzu said:

> *Hence a general is skillful in attack*
> *whose opponent does not know what to*
> *defend; and he is skillful in defense*
> *whose opponent does not know what to*

attack. So the art of subtlety and
secrecy! Through it we learn to be
invisible, through it inaudible; and
hence we hold the enemy's fate in our
hands.

So important is the art of subtlety and secrecy that Sun Tzu also recommended taking severe measures against those unauthorized individuals, enemy or friend, who learn about the mechanism of a mystery while that mystery remains necessary. Sun Tzu said:

If a spy divulges a secret piece of news
before the time is ripe, he must be put to
death together with the man to whom he
told the secret.

The matter of secrecy inherent in mystery has serious implications for those desiring efficient victory. Bottom line: keep your secrets secret.

Now, having used maneuver and positioning to enhance and leverage deceptions, you act to secure your victory. On that we have Sun Tzu's fourth principle, energy.

4 - Energy

To achieve victory in war you take effective action against your enemy. The force applied against your enemy during that action determines the immediate impact of that action and so becomes a means to predict that action's effectiveness. Momentum determines the sustainability of an action and figures directly into the calculation of force. Flow determines where and how an action delivers force. The total force applied to the target of an action upon impact follows the line of the physics equation:

$$\text{Force} = (\text{mass x velocity}) / \text{time}$$

Energy in war represents a combination of all the variables represented in the equation above. It determines both the ability of the offense to apply the required force on an objective necessary to succeed and the ability of the defense to oppose or avoid that force. The most severe application of force in war developed from Albert Einstein's equation:

$$E = mc^2 \text{ or } (\text{Energy} = \text{mass} * \text{constant}^2)$$

This equation above became the foundation upon which nuclear physicists built thermonuclear bombs. Little can oppose the force these weapons deliver.

In the natural world, conservation of energy represents the normal state of affairs. As Sir Isaac Newton described, an object at rest will stay at rest, and an object in motion will stay in motion, when no other force asserts itself on those objects. With regard to living entities, where more efficient ways to use energy evolve over time, those who expend less energy to obtain the same result tend, over time, to displace those who expend more. This truth applies even to those who would use nuclear weapons in dual fashion. First, those who can win conflicts without resorting to nuclear weapons preclude massive collateral damage to anything near a target. Collateral damage wastes energy, serves no purpose for the attack, and can even hinder the greater purpose behind an attack. Second, one nuclear armed delivery system can apply as much force on a target as dozens of conventional delivery systems and so becomes an efficient way to deliver a massive force if needed. On such nuclear devices, however, we expect that Sun Tzu had no concept. So we look for a more present and smaller scale illustration of efficient energy use that Sun Tzu did observe when he described the application of energy.

THE EFFICIENT FALCON

The falcon on the hunt, for example, spreads its wings wide and soars about in search of prey with little effort, particularly in contrast to smaller birds, sparrows and the like, that must constantly flap their wings to keep aloft. When the soaring falcon finds its prey and dives to attack, it folds its wings into its streamlined body and allows gravity to build the speed and momentum it will use to strike. Also, and like most predators, when the falcon commits to the attack and dives, it also seeks to surprise its prey to avoid having to chase it down. The chase expends unnecessary energy, risks the possibility that the prey might fight back, and overall reduces the falcon's chance of a success – hence the potential to waste energy for no gain. So a warrior should consider the falcon's ways in his own efforts and ask how he can strike successfully with the least expenditure of men and materials. In reference to the falcon, Sun Tzu said:

> *The onset of troops is like the rush of a water torrent which will even roll stones along in its course.*

> *The quality of decision is like the well-timed swoop of a falcon which enables it to strike and destroy its victim.*

OFFENSIVE AND DEFENSIVE ENERGY

In physics, the meeting of opposing offensive and defensive energy produces straight forward and predictable results. The weaker of the two forces succumbs to the stronger. Take the following quote from *The Man of La Mancha* as an illustration of this truth:

> *Whether the pitcher hits the stone or the*
> *stone hits the pitcher, it's going to be bad*
> *for the pitcher.*

Contrast the pitcher and the stone to Sun Tzu's statement that deals with some of the very practical aspects of war and its results. Sun Tzu said:

> *Other conditions being equal, if one*
> *force is hurled against another ten times*
> *its size, the result will be the flight of*
> *the former.*

Unlike the case of the pitcher and the stone, the energy used in association with defense by entities that monitor their surroundings varies with the activity and the threat perceived by the defender, which in turn becomes the basis to make conditions unequal. When the defense opposes an attack, the defense relies upon some combination of strength, armor, speed, or maneuver to thwart the attacker, or as noted in previous chapters, relies on having the foresight to avoid danger in the first place. An oblivious defender can only rely on passive defenses when under attack, meaning perhaps its armor or obscuring camouflage. The oblivious defender therefore opposes an attacker's energy with less defensive energy than if that defender also deployed active defenses and, for example, took cover and fired back. Effective deceptions that keep a present enemy ignorant of an impending attack become a key to conserving energy for an attacker and a key to amplifying the force applied against the defender – and the reverse in principle holds true for the defender. So, on the heels of his opening statement that all war is based upon deception, and on the heels of his statement on the need to appear the opposite of able or unable, active or inactive, and closer or further than you actually are, Sun Tzu continued:

> *Hold out baits to entice the enemy to*
> *act. Feign disorder, and strike him*
> *when he seeks to take advantage.*
>
> *If your enemy is secure at all points,*
> *prepare for his attack. If he has*
> *superior strength, evade him.*

If your enemy bares a short temper,
seek to irritate him. Pretend to be
weak, so he becomes arrogant.

If your enemy takes his ease, give him
no rest. If his forces unite, separate
them.

Attack your enemy where he is
unprepared, appear where he does not
expect you.

These military deceptions that bring
victory must not be revealed as
deceptions before they succeed.

All these recommendations serve to put an enemy at a disadvantage at the critical moment of attack and amplify the effectiveness of the energy you apply against him. Therefore, to apply energy with the maximum effect, do the following:

How to Apply Energy

1. **Direct an appropriate force on your target at the appropriate time where it will have the maximum intended effect on the enemy**
2. **Employ deception to enhance the element of surprise and the probability that your enemy will behave in the way it best serves you**

Applying energy in this way produces dual benefits. First, you simply increase the chance you will succeed. Second, you succeed while using the minimum force necessary. Succeeding with the minimum force necessary means you identify critical points when you attack, like a bridge over a river, a road junction, or a key power plant, where control or destruction of a comparatively small target has a great effect on an enemy's continued ability to fight. Case in point, a stated mission might involve destroying an enemy's bridges from the air, but the greater intended effect of that same mission might involve cutting the flow of supplies that the enemy needs to fight your soldiers on the ground. When determining the appropriate force to use in an attack, think always of the greater intended effect when you plan your missions. You do not, in the above case, want to expend the resources necessary to destroy two bridges along the same highway when destroying only one bridge will cut the flow of supplies along that highway just as well, freeing up the second resource to go after a different target. Also, in regards to the greater intended effect of a mission, allow deception to facilitate surprise and encourage an enemy to position himself where you can have the greatest impact on him or an intended target. Traps rarely work without the right bait, and even tigers on the hunt hide in the forest.

BATTLE OF THE BULGE

On December 16, 1944, seven months after D-Day, German armored divisions that had secretly assembled near the German border smashed through American infantry divisions in the Belgian Ardennes forest. Massive Tiger tanks, belonging to an enemy many had considered all but defeated, rolled down wooded roads considered so unfavorable to armored warfare that the allies had sparsely defended them. Americans experienced the power of a German blitzkrieg for the first time.

Blitzkrieg means "lightning war." It is a form of maneuver warfare that involved close coordination between air power and mobile land forces. The German's used the aerial element of the blitzkrieg to destroy command and control centers in order to put their enemy into chaos and disrupt communications. German armored and mobile infantry forces advanced behind these strikes against a weakened enemy. The tactic models Sun Tzu's philosophy on energy. Sun Tzu said:

> *Therefore, the good fighter will be overwhelming in his assault, and deliberate with his timing.*
>
> *His energy may be likened to the bending of a crossbow; his timing to the release of a bolt by the trigger.*
>
> *Amid the turmoil and tumult of battle, there may appear disorder and yet no real disorder at all; amid confusion and chaos, you may appear as if without head or tail, yet you will be invulnerable to defeat.*

By late 1944, the German's lacked the air power normally associated with a blitzkrieg. However, German commandos dressed in American uniforms and fluent in American slang facilitated, at least psychologically, the disruption of communications and the creation of chaos upon which the blitzkrieg relied. Their existence caused American units to become suspicious of themselves. The German decision to attack during bad weather also deprived Allied commanders of aerial reconnaissance and the means to use Allied air power to disrupt the momentum of the attack. This meant American ground forces faced the German offensive on their own.

When German armored and mobile infantry units advanced, the American line crumbled. American infantry forces, many of them inexperienced or reorganizing, could not stand up to some of the best divisions and most seasoned troops that remained in the German military. To the German advantage, the Americans fled after receiving the initial impact of the assault. Historically it has been very difficult for any routed army to regroup and regain the momentum of a battle as long as its attacker maintained effective pressure.

To the German disadvantage, however, the restrictive terrain in the Ardennes forest made the capture of towns with key road junctions imperative to the success of their mission. These junctions became a means for American soldiers further behind the lines, and therefore not immediately subjected to German fire, to focus their defensive efforts. Holding these junctions became the immediate objective for these American forces. Bringing the German advance to a halt became the greater intended effect that holding these junctions would deliver.

PATTON

America's Gen. George S. Patton, fighting south of Belgium when the Germans attacked that December, had also mastered the principles of energy. Lack of fuel and other critical supplies, not the Germans, had stopped the momentum of his 3rd Army's drive toward Germany some months earlier. That delay afforded the Germans a chance to regroup.

The German high command feared Patton if for no other reason than Patton understood and used the power of blitzkrieg as well as or better than their own generals. Patton used the principles of blitzkrieg against them. He summed up its principles in his May 31, 1944 speech when he said:

> *Our basic plan of operation is to advance and to keep on advancing regardless of whether we have to go over, under, or through the enemy. We are going to go through him like crap through a goose....*

A blitzkrieg did not allow pockets of resistance to stop the flow of an attack. Instead, the blitzkrieg bypassed pockets of resistance and dealt with them later. Surrounded pockets of resistance were worn down or forced to surrender in due course. But like flowing water, an army needs sufficient momentum to overcome certain obstacles, or it stops and dries up. The German offensive against the Americans failed in part because the Germans could not wear down one critical pocket of resistance. The 101st Airborne Division held a key road junction at Bastogne until relieved by Patton's 3rd Army. The German offensive, forced to drive around Bastogne as a result, soon lost the momentum it needed to reach its objective, the critical Allied supply port of Antwerp. Had they reached Antwerp, the Germans had intended to cut off supplies critical to the Allied march on Germany.

IN BALANCE

In balance, then, to fight efficiently yet still succeed, you want your initial force to lead to a decisive victory at the outset, or if not, then you want the momentum of your initial impact to carry you to victory shortly thereafter. Like a bicyclist who peddles fast to gain the momentum to climb a hill, you want your momentum to carry you over the top without too much additional struggle. If, however, your initial force will not secure the victory, then, to fight efficiently, you pace the expenditure of energy for the long haul, in this case being like the bicyclist who climbs a hill at a moderate pace and in low gear so as not to burn out and fail to complete the rest of the course. In either case, flow then determines the additional level of effectiveness for which you apply energy – whether you, as the bicyclist, choose a route over that hill when you can use a tunnel instead. That leads to the importance of Sun Tzu's next principle, strengths and weaknesses, when determining how best to use energy.

5 - Strengths and Weaknesses

So the question arises: how to use energy against an opponent most effectively? To answer this question, Sun Tzu appears to have studied the Taoist philosophy on the Way of life very closely. We can make many direct comparisons between Sun Tzu in *Sun Tzu on the Art of War,* and Lao Tzu in the *Tao Te Ching,* with a consideration that the ideas on the Way of life, described by Lao Tzu in the *Tao Te Ching*, appear to be the primary source for Sun Tzu in *Sun Tzu on the Art of War.* Sun Tzu appears to have used an enemy's violations of the Way of life to ferret out the actual weaknesses to use against that enemy.

Throughout *Sun Tzu on the Art of War*, Sun Tzu made clear the importance of attacking an enemy's weaknesses. That importance to his philosophy, in itself, takes little interpretation. Sun Tzu said:

> *You may advance and be absolutely*
> *irresistible if you attack the enemy's*
> *weak points.*

Though this single idea to avoid strengths and attack weaknesses has probably made it into more strategy planning rooms than any other idea from Sun Tzu, leveraging this idea well requires considerable study.

FINDING EXPLOITABLE WEAKNESSES

Occasionally you can find or create an obvious and exploitable weakness in an opponent. When you do, take the advantage. However, astute opponents tend to compensate for their weaknesses and may even use them to lure you into a trap. Finding an exploitable weakness successfully more often means turning an enemy's apparent strength into a weakness before you begin your decisive engagement. By turning an enemy's apparent strength into weakness, you afford yourself the possible advantages that you attack your enemy's weakness while your enemy believes it represents strength or you attack your enemy's weakness while he is too committed to his position to make an adjustment. As per the principle of deception, you can often use your enemy's needs, desires, fears, and any of the five dangerous faults he may have, to further leverage his inability or unwillingness to compensate for a weakness until well after your attack has succeeded.

To understand how a perceived strength can become a weakness, consider how Col. Billy Mitchell's demonstration in 1921 that an airplane could sink a battleship, and its resultant challenge of the battleship as the Navy's premier weapon, so severely riled Secretary of the Navy Josephus Daniels that it led to Mitchell's court-martial four years later. Consider also how the German belief in the supremacy of their present armored forces through 1941, reinforced by their success at conquering most of Europe, left them without a tank powerful enough to destroy a new Russian tank called the T-34. When the Russians fielded the T-34 in force during a surprise counterattack outside Moscow in December 1941, the best German tanks available could not match its speed, armor, cross country ability, firepower, or reliability. Only direct fire from redeployed 88 millimeter anti-aircraft cannons could reliably destroy the new Russian tanks at long range. So in both these illustrations, commanders perceived they had strength until reality showed them wrong.

Lao Tzu described how strength becomes weakness in the following verse from the *Tao Te Ching*. Note how the weaknesses Lao Tzu described actually derive from what might otherwise be described as strengths:

> *Fill a vessel to the brim, and it will spill.*
> *Keep to sharpening a blade, and it goes*
> *thin and blunt.*
> *Surround yourself with treasure, you*
> *cannot keep it safe.*
> *A proud man falls to his knees.*
> *Wealth and honor lead to arrogance,*
> *which brings evil upon itself.*
> *When the work is done and one becomes*
> *distinguished, to retire is the way of*
> *Heaven.*

Sun Tzu's own verses on strengths and weaknesses in *Sun Tzu on the Art of War* likewise focused on turning apparent enemy strengths into weaknesses. They also served as a warning not to allow your own strengths to become weaknesses. For example, Sun Tzu warned that the extended use of a strong and sharp army in the field makes it dull, like the dulled blade described in Lao Tzu's verse above, and also depletes the treasury. The impact is both practical and psychological. Consider how Russian commanders in the aforementioned scenario deliberately waited until attacking German divisions had fully extended themselves to the very gates of their desired objective, Moscow, before they counterattacked with fresh troops and the T-34s, this to further the psychological impact their attack would have against an exhausted enemy that believed they had almost won. Sun Tzu said:

> *When you engage the enemy in actual fighting, if victory takes a long time to achieve, then your men's weapons will dull and their enthusiasm for the fight will diminish....*

> *...Now, when your weapons are dulled, your enthusiasm diminished, your strength exhausted and your treasure spent, other chieftains will appear to take advantage of your weakened condition.*

The Germans had counted on achieving victory in 1941, before the Russian winter set in, familiar for sure with the fate of Napoleon when he failed to do so in Russia in 1812. Like Napoleon, the failure to achieve victory before the winter effectively cost them the war. Soon the Germans also faced advances from Russia's rival but allied "chieftains," most notably the United States and Great Britain.

TAKING SOMETHING OF VALUE

Sun Tzu also described the opportunity to attack weakness inherent in an enemy burdened to protect something of value. This idea follows along the line of Lao Tzu's statement that a man surrounded by treasure cannot keep it safe. In this idea, another dual point about potential weakness appears. The strength of treasure, which can mean abundance in general, causes weakness when an enemy cannot protect it and can also become weakness if it encourages an enemy to take ill adventures. In either case, Sun Tzu advised military leaders to use an opponent's treasure against him, particularly if treasure forms the basis of his strength, because you threaten his treasure itself or act where and how his possession of treasure does not provide him with an advantage. Sun Tzu said:

> *If asked how to cope with a great number*
> *of the enemy arrayed in orderly fashion*
> *and in the act of marching to the attack,*
> *I should say: "Begin by seizing*
> *something which your opponent holds*
> *dear; then he will be amenable to your*
> *will."*

Sun Tzu advised not to attack a well arrayed enemy – his strong point – but instead advised to threaten something that enemy must protect – his weak point. This common strategy crosses all professional disciplines and means threatening what an opponent holds dear to keep that opponent from taking effective counteraction. Moral considerations and all potential larger consequences set aside, this idea, which is fundamentally the hostage strategy, can work well when a vulnerable opponent can also act on its enemy's demands. In its grandest use to date, the United States and the Soviet Union used the mutually assured destruction of each other's cities to keep the peace and the world whole for nearly 50 years. Though often called by the more neutral name of deterrence, the United States and Soviet Union held everyone within each other's cities hostage to keep the peace. Outside war, if you have ever needed to work at a position you would otherwise leave in order to vest in a retirement plan, if you have ever had a boss threaten your job if you did not accept an unacceptable task, or if you have ever had a criminal threaten your life, then you know how well this strategy can influence your immediate actions.

WATER

To further describe strength and weakness, Lao Tzu, in the *Tao Te Ching*, made many references to the dual nature of water. Lao Tzu said:

> *What is more fluid, more yielding than*
> *water?*
> *Yet back it comes again, wearing down*
> *the rigid strength, which cannot yield to*
> *withstand it.*
> *So it is that the strong are overcome by*
> *the weak.*

In *Sun Tzu on the Art of War*, Sun Tzu also made extensive use of water imagery to describe strengths and weaknesses. His readers, familiar with water as a key source of life in an agricultural society – and a key threat at times – could easily understand this model. Sun Tzu said:

Tactics are like unto water; for water in
its natural course runs away from high
places and hastens downward. So in war,
the way to fight is to avoid what is strong
and attack what is weak.

The importance of these passages for understanding strengths and weaknesses becomes apparent when taking action. Like water, strengths and weaknesses have a dual nature and depend upon the circumstances of their condition. Therefore, to make strengths and weaknesses work for you, remember the dual nature of all advantages and disadvantages, and do one or more of the following:

How to Use Strengths and Weaknesses

1. **Use your enemy's inherent or situational weaknesses against him**
2. **Turn your enemy's apparent strengths against him**
3. **Turn your apparent weaknesses into strengths**
4. **Use stratagem to mask weaknesses and exploit strengths**

Think about these statements figuratively in terms of the river described at the beginning of this book. If your opponent takes his stand against you on a foundation of sand, by all means use the inherent weakness of both his situation and that foundation against him. Wash the sand away. If your opponent hides in the apparent safety of a solid fortification and you can flood him out, then do that. The longer he continues to believe in the strength of his fortification, the better. If the flow of your river cannot immediately wash away his fortification, then perhaps you can enlist time as an ally and wait for your river to wear down his fortification. Last, if you just threaten to flood or wait for your river to wear his fortification down over time, even if only a bluff, perhaps your opponent will surrender before you must see your threat through.

Therefore, consider any action you take in line with how a river journeys to the sea. Ultimately, your best way to the sea follows the path of least resistance. If you hit an obstacle along your way and cannot go through it, then go around or over it. Consider using a bluff if it can facilitate your way through and you can accept the consequences of its possible failure. Whatever you do, do not stop in your quest to reach your goal. If you stop, you stagnate and dry up. Deviate from the path of least resistance only to the point you do not allow it to channel you into a dead end.

Of course, with enough power behind you, you can always travel against the flow of a river for a time. Do so with care. If you have the propensity to do this often, remember the way of the salmon. It may indeed complete its journey upstream, but then it dies.

GRANT

In May of 1864, ten months after Gettysburg, when 120,000 Union soldiers under Gen. Ulysses S. Grant attacked Gen. Robert E. Lee's 60,000 Confederate soldiers at the battle of the Wilderness, Grant knew that having a superior number of troops on that battlefield did not afford his real strength against Lee. His real strength was that no matter the losses he bore, he could replace his casualties and Lee could not. In effect, he could force Lee to become like a boulder in a river that his impending flood of Union soldiers would wear down.

Facing a numerical disadvantage, Lee predictably built defensive positions in difficult terrain where Grant could not bring the full power of his larger army to bear on any decisive place. Lee's soldiers then proceeded to inflicted huge casualties on Grant's army. Grant expected and accepted this result. Then, after the battle, rather than allowing Lee's defenses to halt the Union's advance, Grant continued south to force additional engagements and wear down Lee's army.

Though Lee won the Battle of the Wilderness on paper, and many battles that followed, he sustained irreplaceable casualties that cost him more than Grant's replaceable casualties cost Grant. Lee's army had fewer than 10,000 soldiers remaining when he surrendered at Appomattox a year later. Though many historians criticize the bloody nature of Grant's methods, under the circumstances, it proved a price his nation was willing and able to bear to keep America whole.

STRENGTHS FROM WEAKNESSES

In *Sun Tzu on the Art of War*, Sun Tzu wrote some very practical messages about strengths and weaknesses with regards to battlefield tactics. For example, Sun Tzu accounted for the well known fact that attacking up hill requires more energy than attacking downhill and exposes attacking soldiers to overhead fire. Sun Tzu said:

> *It is a military principle not to advance uphill against the enemy, nor to oppose him when he comes downhill.*

The real power of Sun Tzu, as always, lies in the subtlety.

To understand Sun Tzu's subtlety in *Sun Tzu on the Art of War*, we again turn to Lao Tzu in the *Tao Te Ching* for an explanation. Lao Tzu wrote how weakness becomes strength when he said:

> *Yield and you need not break:*
> *Bent you can straighten;*
> *Emptied you can hold;*
> *Torn you can mend;*

Many people who reference Lao Tzu's verses, like the verse above, tell how a strong tree will break in a storm while a weak blade of grass bends in the wind and survives. They also describe how a living blade of grass stays supple and a dead blade of grass becomes hard and brittle, all with the idea, literally and figuratively, that life and flexibility lead to strength and death and rigidity to weakness. Along a similar line, Sun Tzu said:

> *To refrain from intercepting an enemy*
> *whose banners are in perfect order, to*
> *refrain from attacking an army drawn*
> *up in a calm and confident array: — this*
> *is the art of studying circumstances.*

To understand the relation between the two passages, understand that to refrain from attacking requires bending somewhere without breaking and requires maneuvers that doctrinal rigidity does not always favor. Such action may mean trading vast expanses of land until environmental conditions wear an enemy out, as the Russians did to Napoleon in 1812, and later to the Germans in the 1940s. It may mean waging a protracted campaign like the American Continental Army fought when it defeated an England concerned with an even greater military threat from France.

During the American Revolution, Gen. George Washington learned from his early defeats that he could not win fixed battles against superior British forces. He therefore informed the Continental Congress that he planned to change his strategy entirely. He would avoid head-to-head battles with the British and seek a protracted war instead, effectively causing the British to violate Sun Tzu's aforementioned observation that:

> *No country has ever benefited from a*
> *prolonged war.*

By yielding on the fixed battlefield, where British soldiers reigned supreme, Washington saved his own army from destruction until changing circumstances, and assistance from France, allowed for a decisive American victory at Yorktown.

ROADS NOT TO FOLLOW

Sometimes an analysis of strengths and weaknesses may show an insurmountable weakness in your own position. Better to know that before a military engagement than after it. Even great courage cannot compensate for a reality that a fight will bring your defeat. So Sun Tzu said:

> *There are roads that must not be followed, armies that must not be attacked, cities that must not be besieged, positions that must not be contested, orders from the sovereign that must not be obeyed.*

Fighting a battle you cannot win is worse than not fighting at all, for in doing the former you also lose your army and your opportunity to achieve victory elsewhere. When under pressure to take action, showing patience until a better opportunity arises or can be created may take more courage from a commander than immediate fighting; however, that patience can prevent disasters.

During the American Civil War, for example, many of Gen. Robert E. Lee's battlefield successes occurred when he incited Union commanders to assault prepared Confederate entrenchments. In fact, for both sides, the military unit actually making an advance across open ground tended to suffer severe casualties when compared to the defending unit it faced, particularly when that defending unit fought from behind or within a prepared position. The accuracy of rifled muskets used by both sides in the American Civil War made any given soldier on open ground an easy target.

Many of Lee's successes when repelling Union advances occurred when he deployed his soldiers into defensive positions that bore an offensive nature. Such positions meant Union forces invading Virginia could not ignore Lee's army when they attempted to reach their objective, Richmond, without exposing their supply lines, and Washington itself, to Lee's army. At Fredericksburg, on December 11-15, 1862, this approach to battle brought a decisive Confederate victory when numerous Union assaults on Lee's prepared positions resulted in 13,000 Union casualties as compared to 5,000 of Lee's own.

For those taking the comparative role of the Union in such scenarios, the situation presents a real dilemma. Orders to take "roads that must not be followed" lead to a loss of the command if disobeyed and a loss of life if obeyed. Like so much about strategy that Sun Tzu discussed, the scenario plays out time and again in other professions where the vast majority of people choose to protect their livelihoods, consequences be damned. They in effect take their charges, be they soldiers, the careers of others, or the wealth of an organization, and carry them across open ground in a dance of death akin – in principle and with all due respect – to that experienced by thousands of Union soldiers Lee's soldiers cut down at Fredericksburg during repeated and hopeless assaults on Confederate positions on Marye's Heights. The apparent nobility of that futile Union effort, launched on December 13, 1862, caused Lee to say:

> *It is well that war is so terrible, lest we*
> *grow too fond of it.*

To have an opponent say such words about your efforts serves as little condolence to a failure. In the long run, it pays to have the inner courage of those who do not take these "roads not to follow," yet somehow succeed anyway. Therefore, and if you have the authority to do so, when faced with orders to pursue battles you cannot win, find, obtain acceptance, and use alternative ways to succeed at the greater goal. Such alternative ways should keep those in your charge whole and alive to see a day of victory when circumstances change and wiser heads prevail. Should you not have the authority to veto a directive to take a "road not to travel," and you do not have the option or desire to leave or disobey your orders, Sun Tzu's sixth principle becomes particularly important to you. Sun Tzu's sixth principle, initiative, enhances all of Sun Tzu's other principles. It also serves as the best way to minimize the harmful effect of mistakes.

6 - Initiative

On initiative, Sun Tzu said:

> *Whoever is first on the field and awaits the arrival of his enemy will be fresh for the fight; whoever is second on the field and must hasten into battle will arrive exhausted. Therefore, the clever combatant imposes his will on the enemy and does not allow the enemy's will to be imposed on him.*

Your enemy cannot enact an effective strategy against you if you own the initiative. Even if you have not taken the best course of action to achieve your goal, your hold on the initiative can still preserve your advantage over your enemy. Competition does not always require you to play your best game to win, as long as you somehow play better than your opponent.

Part of taking action to gain the initiative involves accepting the risk associated with acting. Should you or a commander miscalculate the risk, then holding the initiative will still likely produce a better result than if an enemy holds the initiative over you.

CHAMBERLAIN

When faced with yet another attack by Confederate soldiers of the 15th Alabama, under the command of Confederate Col. William Oates – an attack for which his own soldiers had insufficient ammunition left to repel – Union Col. Joshua Chamberlain ordered men in his 20th Maine regiment to fix bayonets at Little Round Top during the Battle of Gettysburg on July 2, 1863. The Union bayonet charge that followed took the initiative away from the Confederate attackers. Surprised by the assault, the Confederates retreated in confusion. With assistance from the 83rd Pennsylvania, Chamberlain's men secured four hundred prisoners from the 15th Alabama and 47th Alabama regiments, including officers. By taking the initiative, and the associated risk, Chamberlain altered a situation that could easily have led to a Union defeat, and thereby secured a Union victory instead.

OWNING THE INITIATIVE

When you own the initiative, your enemy must react to you because you do one or more of the following:

How to Own the Initiative

1. **Threaten to keep your enemy from obtaining his objective**
2. **Threaten to obtain your objective at your enemy's expense**
3. **Threaten your enemy's existence** (often a permanent way to achieve the former two)
4. **Threaten something your enemy values more than his existence**
5. **Employ a deception that leads an enemy to act on a suitable false reality**

Threatening to keep your enemy from obtaining his objective means to prepare a defense that has an offensive nature, and this in turn means that you force your enemy to attack you on ground and under circumstances of your choosing. Threatening to obtain your objective at your enemy's expense means to take the offense and so force him to defend or counterattack, again, on ground and under circumstances of your own choosing. Threatening your enemy's existence means you threaten to destroy him outright with the result that you neutralize his ability to oppose you altogether.

Threatening the existence of something your enemy values more than his existence takes threatening your enemy's existence as an entity one step further. This means that you threaten to neutralize your enemy's opposition, with or without destroying him, and with the added caveat that your hold on that which he values more than his existence prevents him from taking an act, in desperation or otherwise, that could destroy you or your objective. An enemy not subjected to this hold may take a final act of desperation if he feels he has no other choice and may at least prevent you from winning also – or from winning whole. This latter condition is also something to consider when seeking an invulnerable position for yourself since exposure to like action by an enemy proves a foundation for blackmail. The threat falls in line with Sun Tzu's aforementioned quote that:

> If asked how to cope with a great number
> of the enemy arrayed in orderly fashion
> and in the act of marching to the attack,
> I should say: "Begin by seizing
> something which your opponent holds
> dear; then he will be amenable to your
> will."

Expect that an able enemy will try to do to you what you intend to do to him if you allow him the opportunity. He might also be able and willing to take actions you would never take. So it pays to prepare in advance.

With regards to employing a deception that makes an enemy act on a suitably false reality – that perhaps, for example, you can act on a threat when or where you cannot – though it necessarily increases your risk that the enemy will call your bluff, it also enhances your power to succeed with the forces you actually have. It represents a key part of the art of warfare, for once an enemy knows you have a willingness to deceive him, you can encourage him to call your bluff the next time when you actually have the capacity to defeat him. As a comparison, think how much easier it is to play poker with those who only stay in a game when they have a good hand. You can measure your bets accordingly. War is a conflict of knowns and unknowns between enemies that must account for the psychology of commanders and their soldiers. When in the role of a commander, you need to keep both practical and psychological aspects of initiative in mind.

DIRECT AND INDIRECT

To gain the initiative in warfare, Sun Tzu discussed the use of direct and indirect maneuvers. He said:

> In all fighting, use the direct method for
> joining battle and the indirect method
> to secure victory.

Sun Tzu also said:

> *The direct and the indirect attack lead*
> *on to each other in turn. It is like moving*
> *in a circle – you never come to an end.*
> *Who can exhaust the possibilities of their*
> *combination?*

In its purist form, you can observe direct and indirect maneuvers when watching a martial artist fight. A martial artist in singles combat might throw a direct blow with his sword and fully expect that his opponent will block that direct blow, this being the act of joining battle. From that direct blow, he will then throw an indirect blow with his sword that his opponent, now out of position, cannot block, this being the act of securing victory. The idea has many combinations. Sun Tzu said:

> *In battle, there are not more than two*
> *methods of attack – the direct and the*
> *indirect; yet these two in combination*
> *give rise to an endless series of*
> *maneuvers.*

Scale up the indirect attack demonstrated by our martial artist and you find how generals used the classic flanking maneuver to bring overwhelming force against the weak side of an enemy line. Confederate Gen. "Stonewall" Jackson used a classic indirect flanking attack to defeat Union Gen. Joseph Hooker at Chancellorsville, on May 2, 1863. He swept around the Union right from the west while Hooker concerned himself with the direct threat from Gen. Lee on his southern front. Jackson's initial assault so surprised Hooker's Union soldiers on that western flank that Jackson's battle line charged through encampments where Union soldiers rested and cooked their meals. Move forward in time, and though technologies change, the underlying idea stays the same. The Iraqi army learned this the hard way when coalition armored forces swept around Kuwait from the west, in February 1991, and struck their fortifications from behind.

A commander also has psychological and economic warfare to use when setting up an indirect attack. Often, but not exclusively, he deploys these measures on the grand strategic scale. He can use psychological operations to encourage enemy soldiers to surrender, or perhaps gain the support of local opposition to those enemy soldiers before his own soldiers must engage in combat. He can use economic measures to strain an enemy's capacity to wage war. The North Atlantic Treaty Organization (NATO) and its allies, on a grand strategic scale, used military parity in conventional and nuclear weapons as a direct force to contain Soviet expansion. As an indirect force, NATO and its allies leveraged weakness in the Soviet economic system. By the late 1980s, the Soviet Union could not afford an expanded arms race threatened by U.S. President Ronald Reagan, nor could it sustain the war in Afghanistan with a Mujahideen enemy NATO helped sustain. When the Soviet economy collapsed, the Cold War ended with a fraction of the losses an armed and probably nuclear confrontation would have created.

THE IMPORTANCE OF SPEED AND INSIGHT

Indirect attacks work best if an enemy does not expect or cannot react to them. Otherwise, they become direct attacks, and an enemy can adjust his defenses to receive them. Hence, the importance Sun Tzu placed on speed and surprise for indirect attacks. Sun Tzu said:

> *Speed is the essence of war: take advantage of the enemy's unpreparedness, make your way by unexpected routes, and attack unguarded places.*

Sun Tzu further advised the use of traps, bait, and ambushes to foster the initiative. For example, if you can force an enemy to take action by threatening a needed supply line, then you know what he must do to defend himself. If you know what he must do to defend himself, then you can better forecast and take advantage of his actions. If you also know your enemy's desires and fears in addition to his ability, disposition, willingness to act, and any faults he may have, your ability to keep the initiative over him rises even further. Hence, Sun Tzu said:

> *If you know the enemy and know yourself, you need not fear the result of a hundred battles. If you know yourself but not the enemy, for every victory gained you will also suffer a defeat. If you know neither the enemy nor yourself, you will be defeated in every battle.*

Knowledge of yourself and enemies must first emphasize needs and desires. Needs and desires forecast both the actions you will take and the actions others will take. Fear, ability, disposition, the willingness to act, and any given faults, determine how and whether you or an enemy will act on those needs or desires. Know these in yourself and in your enemy and you know how and whether to commit to a battle. When you know how and whether to commit to battle, your chances of achieving victory rise significantly. When you can further prevent an enemy from knowing you, you can make your victory complete.

DIFFICULT POSITIONS

However, whether to fight is not always an option, especially if an enemy chooses to attack you. So what then do you do when you know the odds in battle clearly favor your enemy, yet you must fight? The time may come when you wish to avoid enemy contact because your weaknesses will not allow you to gain or keep the initiative in battle, yet you dare not or cannot retreat to a safe haven. This situation can occur when defending and can also occur when attacking, particularly during times of temporary vulnerability, such as when a force in motion has not had time to establish itself in a new position. On this, Sun Tzu said:

> *If we do not wish to fight, we can prevent the enemy from engaging us even though the lines of our encampment have no fortified defenses. All we need to do is to throw something odd and unaccountable in his way.*

More often than not, this unaccountable thing is a deception designed to last at least long enough to keep the enemy from taking the initiative.

D-DAY

On June 6, 1944, during the Allied invasion of Normandy on D-Day, the German high command kept on reserve a number of armored divisions they might otherwise have used against weak allied beachheads before those beachheads became established. Gen. Dwight D. Eisenhower confused the German high command by keeping Gen. George S. Patton out of the Normandy invasion force.

Instead of using Patton in the actual invasion, Eisenhower created a fictional army under Patton's command in an operation called "Fortitude South." This operation came complete with dummy landing craft and inflatable tanks for German reconnaissance planes to photograph, plus false radio traffic to simulate an appropriately sized army. The German high command expected the Allies to use their boldest general to lead the invasion across the English Channel, and Patton's assignment to this fictional unit showed the German command what they expected to see. Furthermore, allied code breakers in Bletchley Park, England, knew from decoded German messages that their deception was still working when D-Day began and so informed the Allied command of this success. By the time the German high command understood that the Normandy invasion was not a diversion, the Allies had established their beachheads and begun their drive into France. The rest is history.

Observations of Sun Tzu's Principles Outside the Military

Sun Tzu did not invent the principles recorded in *Sun Tzu on the Art of War*. He observed them – used them in accord with Chinese history – and then presented them in a way that made sense to his followers. Actions consistent with Sun Tzu's principles continue to take place every day, clearly visible for those who watch. The pure logic of Sun Tzu's principles foregoes any requirement that those applying them actually read *Sun Tzu on the Art of War* to have discovered them on their own. For an examination of Sun Tzu's principles outside the military, start with his first principle, winning whole, and, as an illustration, suppose Sun Tzu practiced his art in business instead of war.

ON WINNING WHOLE

If Sun Tzu had practiced his art in business, he might have approached the idea of winning whole like this:

In the practice of the art of business, the best way to earn a market is to earn it with your profitability intact. To earn a market while destroying its profitability is inferior to this way.

Hence, to win every contested sale through skillful negotiation is not the foremost excellence. To sell to a buyer without a competitive contest is the foremost excellence.

These statements above mean sell what, how, or where a competitor does not sell in order to command high profits. Limit a buyer's decision to include only whether to buy your product, and preclude any need for that buyer to decide from whom to buy your product. Do not compete on price alone. So, for example, if you sell something buyers want that only you can offer, you have more power to earn higher profits than otherwise. Consider the profits the Microsoft Corporation has generated with its exclusive control of Windows software as an illustration. If you also establish a strong relationship with a buyer and charge a fair price, that buyer might never invite a competitor to challenge you. With a strong enough relationship in place, a buyer may hesitate to offend you with such an invitation to a competitor.

In other professions, the principle of winning whole also applies. Carry winning whole into the legal profession, and the classic question of whether to settle a lawsuit or pay to fight in court takes hold. Carry winning whole into medicine, and the question of how to beat a disease without also killing the patient takes hold. In sports, a professional athlete who needs to win a present game must also concern himself to stay fit enough to play the next games after – this a major consideration for Michael Jordan when he came out of retirement, in 2001, to play for the Washington Wizards. In life, the question of how to have a successful career yet still have a fulfilling personal life arises. The principle of winning whole addresses all these issues and helps to outline the ideal solution.

ON LEADING TO ADVANTAGE

Answering the question of how to win whole starts with Sun Tzu's second principle, leading to advantage. This principle in part translates to "stack the deck in your favor" since by stacking the deck in your favor you do and can lead others to further advantage. In business, unique knowledge, a patent, a contract, and a number of other devices that create barriers to competition provide advantages that can stack the deck in favor of attaining high profitability. In law, irrefutable evidence against one party might force a settlement in favor of the other. Having a sound and healthy body stacks the deck in favor of those recovering from physical injury. Superior conditioning allows athletes to push for a present win without jeopardizing their chances to win in the next games after. In life, those who have more to trade for money than just their time – perhaps a unique and valuable talent – better their chances of both achieving success and having a rewarding personal life, depending upon how they use their advantage. Consider how actresses like Meryl Streep and Nicole Kidman leveraged their talent and appearance to succeed in the entertainment industry – this also to the advantage of the entertainment industry. On comparable examples, Sun Tzu might have started his discussion by saying:

> *Avail yourself to good fortune and create situations that will lead to success and sustainable advantages.*
>
> *When such favorable circumstances arise, modify your plans accordingly to take advantage of them.*

Stack the deck in your favor whenever you have the opportunity to do so. When an opportunity or threat presents itself, call upon those advantages to win, and to win whole. Those who succeed by using overwhelming advantages can always show restraint after their victory.

ON DECEPTION

The ability to deceive if necessary, and also to detect the same, is a major advantage for those seeking to win whole. Though the word has negative connotations, it only partly deserves that reputation. Many bright people, for example, want to believe that a special car or fine clothes make them somehow better as a person. The sellers of those products can encourage their buyers' self-deceptions to stack the deck in their own favor during sales negotiations and earn high profits – win whole. Such self-deceptions do not necessarily harm anyone, and for those who like fine things, the right car and clothes can genuinely enhance the quality of life.

The profession of law as we know it is grounded upon the interplay between deception and truth. If people told the whole truth and nothing but the truth without bending it even a little, the courts would only concern themselves with the validity of proposed or taken actions. Since deception is a part of human nature, the courts must sift through truths, deceptions, and appearances from differing points of view to find the most probable truth as a whole. When deliberately abused, deceptions lead to the frivolous lawsuits by which a few attorneys have tarnished the whole industry.

Doctors leverage deception positively when they use harmless vaccines to deceive the body into building immunity against a real disease. They often also prescribe drugs that fool the harmful mechanisms of disease into acting harmlessly. In sports, athletes depend on deception to cause their opponents to make mistakes and so open the way to score points. In personal life and love, Sun Tzu could arguably have said:

> *The art of courtship is based on deception.*

Perhaps mystery, not distance, makes the heart grow fonder. But most people, in relationships or otherwise, do not "bare" well when completely exposed. Hence people often feel the need to reveal and conceal themselves selectively to stack the deck in their favor and obtain their whole desires.

ON ENERGY

Sun Tzu's fourth principle on energy practically translates to "action." When an opportunity presents something of value to win whole, you have stacked the deck in your favor to win it whole, and, through subtle or not so subtle deceptions, you have further reduced the power of your opposition to oppose you, you must act on the opportunity to actually succeed.

The principle of energy connects Sun Tzu's other principles and, like all of Sun Tzu's principles, mirrors Sun Tzu's observations about the ways of life. Energy is life, and sustaining life involves taking action from a position of advantage under conditions whereby even when not successful at an endeavor, life itself at least stays whole. Sun Tzu could have said:

> *Therefore, to make yourself secure from defeat while also positioning for advantage, behave like the tiger, hidden but alert under the cover of brush, safe from stampeding herds but ever repositioning, ever ready to chase down an opportunity in an open and public flash.*

Critical moments come and go, and only the most astute and able practitioners of an art capitalize on them where others must stumble. Consider, as a case in point, the actions of one highly successful basketball coach. Coach Mike Krzyzewski has coached far more last second wins than he has losses with the Duke University Blue Devils basketball team. Coach Krzyzewski appears to have an exceptional ability to read the positioning and the psychology of opposing players and can use that to direct the actions of his own players to where they produce a win.

Regarding energy in business, the best salespeople sense the moment when a prospect is ready to sign, and then act to secure the order. In law, the best lawyers position the opposition, judge, and jury with the evidence, and in the critical moment, make their case on an irrefutable idea, not too early, not too late, so the idea has the momentum to carry the victory. The skillful surgeon cuts out the disease yet leaves behind tissue in good health for continued life. The sports team captain calls a play on a quick count to keep up the momentum of a drive down the field. A chance meeting in an elevator leads to a great new job because you already know what you want and can act on the opportunity presented to you. All of these illustrations are principles of energy and a key to taking effective action.

ON STRENGTHS AND WEAKNESSES

All opportunity has opposition, even when that opposition resides within us. Though facing opposition head on can be a learning experience in practice, when it really matters, energy best focused acts where opposition is weak and not where opposition is strong. At the end of the day in business, true business leaders measure their success on their ability to provide wealth and livelihoods to those associated with their enterprises. Businesses succeed most reliably when they follow a lucrative path of low resistance to that success. They sell benefits that buyers want to buy in ways competitors cannot easily match. They sell hard while the opportunity to sell exists and always sell in accord with their advantages.

In business, however, acting in accord with strengths and weaknesses requires an added subtlety over the military. Opposition comes from many angles and rarely denotes a contest between just two sides, as in war. Multiple competitors create multiple types of opposition simultaneously. A business leader's primary objective, therefore, involves earning customer business and rarely involves the deliberate destruction of competitors as in war. Competitor destruction usually occurs as a byproduct of that business leader's own success in the marketplace. In business, Sun Tzu might therefore have described a broader positioning model to ensure a business leader identifies a place within a market to succeed. Maybe Sun Tzu would have included the following about differentiation:

> *In accord with your strengths and weaknesses, do the familiar differently or make the different familiar. If you do the familiar the same way, buyers will not recognize you. If you do the different in a different way, buyers will not understand you.*

This statement means that in business you should seek to sell a recognizable benefit yet also differentiate your offering from that of competitors. By doing so, you become the preferred choice of a market segment, and as a preferred choice, you can better command profitability within that segment. When you can command profitability within a market segment, you can, from a business standpoint, win whole. Note here the reintroduction of Volkswagen's Beetle, in 1998, as a profitable product both different and familiar to buyers at the same time.

Outside business, where contests do occur between just two sides more frequently, Sun Tzu's principle on strengths and weaknesses, as Sun Tzu originally described it, holds clearer. Lawyers seek juries least likely to reject their arguments, or when on contingency, seek cases effectively already won so their opponents will settle quickly. Though lawyers can profit handsomely by doing so, actually going to trial rarely describes a path of low resistance.

Doctors, likewise, seek less invasive ways to cure disease, and a sports team strong in the center and weak on the outside can expect opponents to attack their weak outside. All of Sun Tzu's principles tie together. If the above sports team can deceive its opponent into believing that it actually has strength on the outside and weakness in the center, its opponent may follow a path of most resistance by attacking that strong center.

In life, acting on what is and not on what you would like it to be is a great first step toward finding a path of low resistance. Not everyone can sing, write, sell, or handle numbers; and a talented writer who becomes an average accountant will prove just as frustrated as a talented accountant who becomes an average writer. Following a path of low resistance requires knowledge of yourself and others which is a key to ensuring others do not deceive you and that you do not deceive yourself.

Effectively following a path of low resistance in accord with *Sun Tzu on the Art of War* also demands that you maintain the integrity and the spirit of the goal you pursue. It involves finding the best way to achieve the intended goal in its entirety on behalf of yourself and those you serve. Thus Sun Tzu said:

> *Now the general is the pillar of the State:*
> *if the pillar has mastered all points of*
> *war, the State will be strong; if the pillar*
> *is defective, the State will be weak.*

So you could expect that those who would compromise on integrity and the spirit of a goal would also themselves prove weak, and those who can achieve a goal in its entirety prove strong.

ON INITIATIVE

Taking the initiative, Sun Tzu's sixth principle, serves as a great way to use your knowledge of self and others effectively and proves a key way to avoid deceptions that would lead you off your best path. Opposition that must react to you cannot effectively deceive you. It must choose a path you decide for it, react from a position of weakness if you have acted well, react at a disadvantage if you have acted wisely, and will retain only the value of a goal you allow for it. Even where you do not ultimately choose your best path – we are all human and make mistakes – your control of the action might at least create a better journey than that experienced by your opposition.

On initiative in business, Sun Tzu might have said:

> He who earns a market forces his competitors into exhaustion if they hasten to follow his success. So effective has he made his position that competitors expend their resources attempting to catch up instead of creating new threats to his business.

This means that if you choose an action and competitors must follow you, even if you did not choose the best action, you still hold an advantage.

Take initiative into law, and you find its importance in winning lawsuits. Those with the initiative can present evidence in a way it serves best; those without the initiative will have difficulty winning even when they should. Sun Tzu, as a trial attorney, might have said:

> Whoever is first on a point of contention and awaits the response of his opponent will be fresh for the fight; whoever is second on a point of contention and has hastened to respond is exhausted.
>
> Therefore, the clever attorney imposes his will on the opposition and does not allow the opposition to impose its will on him.

This means a lawyer prepares himself, the jury, the opposition, and the judge to receive the highest intended impact of any evidence he brings into the courtroom. His success in turn is a measure of:

1. The whole impact intended from the evidence
2. The specific advantage the evidence brings
3. Managing what else is revealed and concealed before the critical moment
4. Acting at the critical moment
5. Acting where the opposition, the jury, and the judge's ability to preclude the evidence is weakest

Those who fight disease must also take the initiative against disease to prevent it from spreading or developing resistance to treatments. Researchers must stay one step ahead of infectious disease by inventing new treatments before old treatments become ineffective, and doctors must facilitate their patient's proper use of the drugs they prescribe in order to prevent the evolution of resistant strains.

Sports teams likewise need the initiative to create more scoring opportunities and to prevent an opponent from scoring in turn. Those who control the ball tend to put more points on the score board.

Our own day also goes better when we, and not others, have control of our schedule. This may mean imposing our schedule on others before they impose it on us.

Bottom line: taking the initiative affords the best opportunity to act on our desires and meet our challenges well – and as a side benefit – offers the best way to compensate for our mistakes.

The Six Sun Tzu Principles as One

In the movie *The Empire Strikes Back,* the second movie of the first released *Star Wars* trilogy, Yoda said to a discouraged Luke Skywalker:

Try not. Do. Or do not. There is no try.

These ten worlds, which paraphrase an ancient Chinese proverb, summarize the heart of *Sun Tzu on the Art of War.* Why is there no try? Because you have already succeeded before the battle begins. To try would mean to seek victory in the midst of a fight, which Sun Tzu said is a sure way to fail. And why do or do not? Because you act when there is something of value you can gain; or you do not act. War offers no middle ground for action that does not pose great jeopardy to the participants and their nation.

Three broad steps, therefore, encapsulate Sun Tzu's principles for all endeavors. These three steps are:

Assess – Prepare – Act

Each of these three steps, expanded below, occur continuously in conflict since they pertain to life and life changes constantly. Their driving force is always a need or desire. All actions, conscious or otherwise, and for survival or for a better life, must begin with a need or desire. All conflict stems from opposing needs and desires. So when faced with an opportunity or a threat, determine the outcome you need or desire and know the outcome your opponent needs or desires. Ask if your need or desire is worth fighting for, assess if you can actually succeed, and make the comparable assessment for your opponent. Every opportunity has a cost, and you must be able to bear that cost, in victory or defeat; and this holds equally true for your opponent. If you can bear the cost in victory or defeat, or should the circumstances created by your opponent require that you fight regardless, plan for and take the three steps below with conviction and focus:

ASSESS

The first step is to assess. The questions below follow the line of Sun Tzu's six principles in the order this book presented them. Any plan you make should include answers to these six questions.

Assess the Opportunity or Threat

1. What do I want to have happen, and what is the whole victory therefore?
2. What inherent and fortuitous advantages do I have, and how can I enhance them?
3. How can I mask my intent from those who would oppose me?
4. What resources do I have and need, and how do I secure my access to them?
5. What effective paths to success lie open to me, and which appear to offer the least resistance to that success?
6. How will I take and keep the initiative so I control the action?

As a result of your assessment, you will do one of two things. You will prepare to act or you will prepare to not act. Your overriding concern involves whether you have something to gain by taking action or whether you have any other choice but to prepare for what conflict may come. You can choose not to throw a rock at an enemy if you cannot bear the consequences of that action, but you cannot wisely choose to ignore that your enemy could or has thrown a rock at you. In either case, when you have completed your assessment, you move to the next step.

PREPARE

The next step is to prepare. Regarding preparation, your plan should account for the following six points. You will put these six points into effect before a decisive engagement with your opponent. Keep in mind that you have achieved your best effect through preparation when your preparation causes your opponent to quit before the actual fight.

Prepare the Field

1. **Affect the conditions of morale that favor your success**
2. **Secure a position of invulnerability**
3. **Assemble the means, the skills, and the authority to succeed**
4. **Deploy and secure the elements of mystery**
5. **Deploy the means to obtain and use foreknowledge**
6. **Monitor the situation for changes in the opportunity or threat**

Bottom line: preparation, not action, is where you win or lose conflicts whole, regardless of who launches an attack. It involves both physical preparation and psychological preparation. From your efforts to prepare for a conflict, you will do one of two things. You will either act or you will not act. You will never, however, link preparation to a commitment for action. At the end of the day, despite all the preparations you might make, you do not jump from an airplane into gale force winds if fortune brings gale force winds onto your drop zone on your chosen day. Such action brings about both defeat and loss. Instead, you reassess and prepare again until an acceptable opportunity presents itself or until you really have no other option but to act. On the latter point, keep in mind – literally and figuratively – only dead soldiers have no means to resist.

ACTION

The third step is action. Action is the focal point of the mission and corresponds with a decisive or the decisive engagement in a conflict. You take action when you have completed your assessments and preparations and pursue a specific objective the attainment of which advances your cause or completes your victory. Even where you must act on an opportunity or threat in a fleeting second, you take action successfully because you have physically and psychologically prepared to act in advance and have obtained the authority to act in advance. In the course of action, Sun Tzu's six principles reverse themselves from the order presented in this book. You started your assessments by defining the whole victory and ended your assessments with how you will control the action, and hence, own the initiative. Then, you prepared to act by securing greater advantages for yourself. Now, in step three, you act as follows:

Take Effective and Efficient Action

1. **Take the initiative to control the action**
2. **Choose the best path to the objective**
3. **Focus your energy on the chosen path**
4. **Deceive your opposition to further clear the way**
5. **Deploy all your advantages to make traveling the chosen path easier**
6. **Obtain the objective intact, destroying neither it nor yourself in the process**

Wait to act until you have effectively secured victory through preparation or until waiting to act brings consequences worse than not acting. In either case, take advantage of an enemy where he has not sufficiently prepared himself. Then, when you engage the enemy, follow Sun Tzu's advice:

> *...when you move, fall upon your enemy like a thunderbolt.*

Where assessment and preparation are times for contemplation and patience, action is not. The longer an action takes to succeed, the greater the negative influence fortune and your opponent may have on that action. If you have made your assessments and preparations well, then you do not rely on good fortune to succeed once your action commences. Fortune, however, always reserves the right to work against you, enhanced if misfortune aids the cause of a yet undefeated opponent. Bottom line: any changes in circumstances over time are more likely to harm your efforts than help them. If you depend on having fair weather for too long, your clear skies will turn to rain. Therefore win and win fast.

Should you find, despite your best assessments and best preparations, that you made a mistake by acting, do not linger at the place of that action. Action, by its nature, expends energy and exposes you to risk. If you determine an action will not deliver the results you desire, withdraw from the action if possible and reassess the situation. But here you have a caveat. What proves good for the individual does not always prove good for the army. It may fall upon a soldier to act against his instincts for self-preservation to support the greater good. In that lies some of the most difficult decisions a soldier in the field can face. The soldier must ask what he and his commander wish and what the desired end is worth? If he answers that question well, and his commander offers the right guidance and purpose, then he will serve the needs of the many that depend on him. In commentary on these difficult decisions, we move to our conclusion of *Understanding Sun Tzu on the Art of War.*

Conclusion

May 3, 1915, after the Second Battle of Ypres in Belgium, a Canadian Lt. Col. by the name of John McCrae wrote a poem:

> *In Flanders fields the poppies blow*
> *Between the crosses, row on row,*
> *That mark our place; and in the sky*
> *The lark, still bravely singing, fly*
> *Scarce heard amid the guns below.*
> *We are the Dead. Short days ago*
> *We lived, felt dawn, saw sunset glow,*
> *Loved, and were loved, and now we lie*
> *In Flanders fields.*
> *Take up our quarrel with the foe:*
> *To you from failing hands we throw*
> *The torch; be yours to hold it high.*
> *If ye break faith with us who die*
> *We shall not sleep, though poppies grow*
> *In Flanders fields.*

McCrae's poem, written from a soldier's point of view, became popular among soldiers fighting in the World War I trenches. It encouraged those who followed the dead into battle to pick up the torch and stay true to the cause of the war lest those former soldiers' deaths be in vain. To the leaders of armies, men such as these still living are the souls of their responsibility, not that they be kept from harm if necessary, but certainly that they not have their lives wasted on battlefields not yet prepared for victory. On war itself, Lao Tzu said:

> *He who serves a ruler of men in*
> *accordance with the way of life will not*
> *subdue a kingdom through force of arms.*
> *Such a course is bound to bring*
> *retribution.*
> *Where troops are stationed, briars and*
> *thorns spring up.*
> *In the track of armies must follow lean*
> *years.*
> *The good general wins a victory and then*
> *stops; he will not continue with needless*
> *acts of violence.*
> *Winning, he does not boast; he does not*
> *triumph; he shows no glory.*
> *He wins because he must.*
> *After winning he will not be overbearing.*
> *Power leads to fall, which is not in*
> *accord with the Way.*
> *What is not in accord with the Way will*
> *surely come to an end.*

In keeping with the Way of life to which Lao Tzu referred, Sun Tzu also wrote about seeking victory in war without seeking glory. Sun Tzu further praised the commander who could achieve victory in war without actual fighting when, in the twice aforementioned verse, he said:

> *Hence to fight and win in all your battles*
> *is not the foremost excellence; to break*
> *the enemy's resistance without fighting*
> *is the foremost excellence.*

Insightful adherence to the idea of achieving victory without fighting, or if that was not possible, fighting without seeking glory has led to great successes even beyond the point of winning wars. Review Lao Tzu's verse above in context with actions Gen. Douglas McArthur took in Japan in 1945. McArthur, on behalf of the United States, accepted Japan's unconditional surrender that ended World War II, but allowed Japan to keep its emperor. This decision proved a key to keeping the soul of Japan whole during the rebuilding that followed. The United States had achieved its intended military victory when Japan surrendered, and McArthur took that achievement no further than necessary. He committed no needless act of violence against the psychology of Japan once the physical contest ended. The foresight of that decision marked an insightful victor and helped secure a peace that economic ties between the United States and Japan keep strong.

CLOSE

War can teach society many lessons for use on and off the battlefield, but the lessons come at a high price to individual lives. Using these lessons to achieve positive results may prove the best gift anyone can offer those individuals who war has touched. Though war may represent the ultimate conflict, all manner of difficulty a soldier faces that hinders his ability to fight an enemy – the unbridled power, narrow thinking, political maneuvering, disease, ineptitude, and apathy that influence war – also influences the day to day struggles in life. In times of peace, these difficulties can lead to war. Sun Tzu's concluding words on war in *Sun Tzu on the Art of War,* therefore, need no explanation. As with so many of his philosophies, they easily translate to other aspects of life: Sun Tzu said:

> *Unhappy is the fate of the commander who strives to win his battles and succeeds in his attacks without exploiting their success and purpose; for the result is a waste of time and general stagnation. Hence the saying: The enlightened ruler lays his plans well ahead; the good general cultivates his resources. Do not move unless you see an advantage; do not use your troops unless there is something to be gained; do not fight unless the contested position is critical. No ruler should put troops into the field because he is enraged; no general should fight a battle simply out of anger. Advance only if it is to your advantage to advance; if not, stay where you are. In time, anger may change to gladness; vexation may change to contentment. But a kingdom that has once been destroyed can never come again into being; nor can the dead ever be brought back to life. Hence the enlightened ruler is heedful, and the good general full of caution. This is the way to keep a country at peace and an army intact.*

Sun Tzu

on the

Art of War

Translated by Lionel Giles in 1910

Edited Version by Robert Cantrell in 2003

I - Laying Plans

1. Sun Tzu said: The art of war is of vital importance to the State.

2. It is a matter of life and death, a road to safety or to ruin. Therefore, it is a subject that must be thoroughly studied.

3. Five constant factors govern the art of war, to be assessed when determining the conditions in the field.

4. These constant factors are:

(1)   The Moral Law
(2)   Heaven
(3)   Earth
(4)   The Commander
(5)   Method and discipline

5, 6. *The Moral Law* causes people to be in complete accord with their ruler and to follow him regardless of any danger to their lives.

7. *Heaven* signifies night and day, cold and heat, times and seasons.

8. *Earth* comprises near and far distances; dangerous and secure positions; open ground and narrow passes; the chances of life and death.

9. *The Commander* stands for virtues of wisdom, sincerity, benevolence, courage and strictness.

10. *Method and discipline* describes the disposition of the army, its subdivisions, the gradation and deployments of its officers, the maintenance of roads that carry supplies to the army, and the management of military expenditure.

11. These five constant factors should be familiar to every general. He who knows them will be victorious; he who does not know them will fail.

12. Therefore, when making your assessment to determine which of two sides has the advantage, ask:

13. (1)   Which of the two sovereigns commands the Moral law?
(2)   Which of the two generals has the most ability?
(3)   To whom lie the advantages derived from Heaven and Earth?
(4)   On which side is discipline most rigorously enforced?
(5)   Which army is stronger?
(6)   Which side has the most highly trained officers and men?
(7)   Which army serves rewards and punishments most consistently?

14. By means of these assessments I can forecast victory or defeat.

15. The general that heeds my counsel and makes these assessments will achieve victory. Let this general be retained in command! The general that does not heed my counsel and makes few assessments will suffer defeat. Let this general be dismissed!

16. While heeding the merit of my counsel, avail yourself also to any helpful circumstances that give you advantages beyond the ordinary conditions.

17. When such favorable circumstances arise, modify your plans accordingly.

18. All warfare is based on deception.

19. Hence, when able to attack, seem as if unable to attack; when using forces actively, seem inactive; when nearby, make the enemy believe you are far away; when far away, make the enemy believe you are nearby.

20. Hold out baits to entice the enemy to act. Feign disorder, and strike him when he seeks to take advantage.

21. If your enemy is secure at all points, prepare for his attack. If he has superior strength, evade him.

22. If your enemy bares a short temper, seek to irritate him. Pretend to be weak, so he becomes arrogant.

23. If your enemy takes his ease, give him no rest. If his forces unite, separate them.

24. Attack your enemy where he is unprepared, appear where he does not expect you.

25. These military deceptions that bring victory must not be revealed as deceptions before they succeed.

26. Now the general who wins a battle makes many calculations in his headquarters before he fights a battle. The general who loses a battle makes but few calculations beforehand. Thus do many calculations lead to victory and few calculations to defeat. What need be said of no calculation at all? It is by observing the points of these calculations that I can foresee who is likely to win or lose in battle.

II - Waging War

1. Sun Tzu said: In the conduct of a war, where an army fields a thousand swift chariots, as many heavy chariots, and a hundred thousand mail-clad soldiers, with provisions enough to carry them a thousand *li*, the expenses at home and at the front, including entertainment of guests, small items such as glue and paint, and sums spent on chariots and armor, will reach the total of a thousand ounces of silver per day. Such is the cost of raising an army of 100,000 men.

2. When you engage the enemy in actual fighting, if victory takes a long time to achieve, then your men's weapons will dull and their enthusiasm for the fight will diminish. If you lay siege to a town, you will exhaust your strength.

3. If the campaign is protracted, the resources of the State will not bear the strain.

4. Now, when your weapons are dulled, your enthusiasm diminished, your strength exhausted and your treasure spent, other chieftains will appear to take advantage of your weakened condition. Not even the wisest council will be able to prevent the resulting consequences.

5. Thus, though we have heard of foolishly rushing to war, we have never seen cleverness in war associated with long delays.

6. No country has ever benefited from a prolong war.

7. Only one familiar with the evils of war can thoroughly know the best way fight it.

8. The skillful military leader does not raise a second levy to fund his campaigns; nor does he load his supply-wagons with provisions more than twice.

9. Bring the implements for war with you from home, but forage on the enemy for additional supplies. Thus the army will have food enough for its needs.

10. Contributing to a distant army impoverishes the state treasury. Contributing to maintain an army at a distance in turn causes the people to be impoverished.

11. On the other hand, an army nearby causes prices to go up and provisions to be depleted; and this steals from the people's ability to sustain themselves.

12. When the local population is impoverished and its ability to sustain itself drained away, the people will suffer even as the government must exact more from them.

13, 14. With this loss of substance and exhaustion of strength, the homes of the people will be stripped bare, and three-tenths of their income will be dissipated; while government expenses for broken chariots, worn-out horses, breast-plates and helmets, bows and arrows, spears and shields, protective mantles, draught-oxen and heavy wagons, will amount to four-tenths of its total revenue.

15. So a wise general forages on the enemy instead of his own people. One cartload of an enemy's provisions equals twenty of one's own, and likewise a single ration of food is equivalent to twenty from one's own stores.

16. In order to kill the enemy, the men must be spurred into a fighting rage; that they may find advantage from defeating the enemy, they must also receive a personal reward.

17. Therefore in chariot fighting, when ten or more chariots have been taken, reward those who took the first. Substitute our flags for those of the enemy, use the captured chariots along side our own. Treat captured soldiers well and provide for them.

18. This is called, using the conquered foe to augment your strength.

19. So in war, then, let a victory be your main objective, not the conduct of lengthy and costly campaigns.

20. Thus the military leadership determines the people's fate, and the man who commands the military determines whether the nation shall see peace or peril.

III - Attack by Stratagem

1. Sun Tzu said: In the practice of the art of war, it is best to take the enemy's country whole and intact. To shatter and destroy his country is inferior to this way. So, too, it is better to capture an army intact than to destroy it, better to capture a regiment, a detachment or a company intact than to destroy them.

2. Hence to fight and win in all your battles is not the foremost excellence; to break the enemy's resistance without fighting is the foremost excellence.

3. Thus the highest form of generalship is to defeat the enemy's plans; the next best is to keep the enemy's forces divided; the next best is to attack the enemy's army in the field; and the worst policy is to besiege walled cities.

4. Do not besiege walled cities if you can possibly avoid it. The preparation of siege weaponry; mantlets, movable shelters, and various implements of war; will take three whole months; and the piling up of mounds over against the walls will take another three months.

5. The general, unable to control his rage toward his besieged enemy, will be tempted to launch his men to the assault like swarming ants, with the result that one-third of his men are slain, while the city still remains untaken. Such are the disastrous effects of a siege.

6. Therefore, the skillful leader subdues the enemy's troops without any fighting; he captures their cities without laying siege to them; he overthrows their kingdom without lengthy operations in the field.

7. With his forces intact he will attain supremacy in his domain, and thus, without losing a man, will achieve a total victory. This is the method of attacking by stratagem.

8. As a rule of war, if our forces outnumber the enemy's forces by ten to one, surround him; if by five to one, attack him; if by two to one, divide his army in two.

9. If equally matched, offer battle; if slightly inferior in numbers, avoid his main force; if quite unequal in every way, elude him.

10. Hence, though a small force may fight a gallant and tenacious fight, in the end, the larger force must capture it.

11. Now the general is the pillar of the State: if the pillar has mastered all points of war, the State will be strong; if the pillar is defective, the State will be weak.

12. There are three ways in which a ruler can bring misfortune upon his army:

13. (1) By commanding his army to advance or to retreat, while ignorant of the fact that it cannot obey. This is called hobbling the army.

14. (2) By attempting to govern his army the way he governs his kingdom while ignorant of the conditions that face his army. This causes confusion in minds of the soldier.

15. (3) By employing officers in his army for reasons other than their military ability, ignoring what it takes to be a military leader. This shakes the confidence of the soldiers.

16. When the army is hobbled, the soldiers are confused and their confidence shaken, rival leaders will surely bring trouble. This brings further disorder to the army and eliminates any possibility of victory.

17. Thus there are five essentials for victory:

(1)   He who knows when to fight and when not to fight will win.
(2)   He who knows how to handle both superior and inferior forces will win.

(3)   He whose army is united in purpose throughout all its ranks will win.

(4)   He who prepares himself and waits to take the enemy unprepared will win.

(5)   He who has military capacity and is not interfered with by the sovereign will win.

18. Hence the saying: If you know the enemy and know yourself, you need not fear the result of a hundred battles. If you know yourself but not the enemy, for every victory gained you will also suffer a defeat. If you know neither the enemy nor yourself, you will be defeated in every battle.

IV - Tactical Dispositions

1. Sun Tzu said: The good fighters of old first made themselves invulnerable to defeat, and then waited for an opportunity to defeat their enemy.

2. To make yourself invulnerable to defeat lies in your own hands, but the enemy himself must provide you with the opportunity to defeat him.

3. Thus the good fighter can make himself invulnerable to defeat, but cannot make the enemy vulnerable to defeat.

4. Hence the saying: One may *know* how to defeat an enemy without having the opportunity to *do* so.

5. Invulnerability from defeat implies the use of defensive tactics; the ability to defeat an enemy means taking the offensive.

6. Standing on the defensive indicates insufficient strength; attacking indicates an abundance of strength.

7. The general who is skilled in defense hides in the most secret recesses of the earth; he who is skilled in attack flashes forth from the highest heavens. Thus on the one hand we have the ability to protect ourselves; on the other hand, we have a complete victory.

8. To predict victory when victory is obvious to anyone with average foresight is not the peak of military excellence.

9. Neither is it the peak of military excellence to fight and conquer and have the whole Empire says, "Well done!"

10. To raise an autumn hair is no sign of great strength; to see the sun and moon is no sign of sharp sight; to hear the noise of thunder is no sign of a quick ear.

11. What the ancients called a clever fighter is one who not only wins, but excels at winning with ease.

12. Hence his victories bring him neither reputation for wisdom nor credit for courage.

13. He wins his battles by making no mistakes. Making no mistakes brings certain victory, for it means conquering an enemy that is already defeated.

14. Hence the skillful fighter takes a position that makes his defeat impossible, and does not miss the opportunity to defeat his enemy.

15. Thus it is that in war the victorious strategist seeks battle after the victory has been won, whereas he who is destined to defeat first fights and afterwards looks for victory in the midst of the fight.

16. The masterful leader cultivates the moral law, and strictly adheres to proper methods and discipline; thus it is in his power to control success.

17. In respect to military methods, we have, firstly, Measurement; secondly, Estimation of quantity; thirdly, Calculation; fourthly, Balancing of chances; fifthly, Victory.

18. Measurement owes its existence to Earth; Estimation of quantity to Measurement; Calculation to Estimation of quantity; Balancing of chances to Calculation; and Victory to Balancing of chances.

19. A victorious army opposed to a defeated one is as a pound's weight placed in the scale against a single grain.

20. The onrush of a conquering force is like the bursting of pent-up waters into a chasm a thousand fathoms deep. So much for tactical dispositions.

V - Energy

1. Sun Tzu said: The control of a large force is the same principle as control of a few men: it is merely a question of dividing up their numbers.

2. Fighting with a large army under your command is nowise different from fighting with a small one: it is merely a question of instituting signs and signals.

3. To ensure that your whole army may withstand the brunt of the enemy's attack and remain unshaken, use indirect and direct maneuvers.

4. That the impact of your army may be like a grindstone dashed against an egg, use the science of weak points and strong.

5. In all fighting, use the direct method for joining battle and the indirect method to secure victory.

6. Indirect tactics, efficiently applied, are inexhaustible as Heaven and Earth, unending as the flow of rivers and streams; like the sun and moon, they end but to begin anew; like the four seasons, they pass away to return once more.

7. There are not more than five musical notes, yet the combinations of these five give rise to more melodies than can ever be heard.

8. There are not more than five primary colors, yet in combination they produce more hues than can ever be seen.

9. There are not more than five cardinal tastes, yet their combinations yield more flavors than can ever be tasted.

10. In battle, there are not more than two methods of attack—the direct and the indirect; yet these two in combination give rise to an endless series of maneuvers.

11. The direct and the indirect attack lead on to each other in turn. It is like moving in a circle—you never come to an end. Who can exhaust the possibilities of their combination?

12. The onset of troops is like the rush of a water torrent which will even roll stones along in its course.

13. The quality of decision is like the well-timed swoop of a falcon which enables it to strike and destroy its victim.

14. Therefore, the good fighter will be overwhelming in his assault, and deliberate with his timing.

15. His energy may be likened to the bending of a crossbow; his timing to the release of a bolt by the trigger.

16. Amid the turmoil and tumult of battle, there may appear disorder and yet no real disorder at all; amid confusion and chaos, you may appear as if without head or tail, yet you will be invulnerable to defeat.

17. Simulated disorder arises from perfect discipline; simulated fear arises from courage; simulated weakness arises from strength.

18. Hiding order beneath a cloak of disorder is simply a question of subdivision; concealing courage under a show of timidity presumes a reserve of energy for attack; masking strength with weakness is to be effected by tactical dispositions.

19. Thus one skilled at keeping the enemy on the move maintains deceitful appearances according to that which will cause the enemy to act. He sacrifices something that the enemy will seek to pursue.

20. By holding out baits, he keeps his enemy on the march; then, with a body of chosen men, he lies in wait for him.

21. The clever combatant looks to the effect of combined energy, and does not require too much from individuals. Hence his ability to pick out the right men and focus military force on a favorable situation.

22. When he uses his combined force, his soldiers become like rolling logs or stones. For it is the nature of a log or stone to remain motionless on level ground, and to move when on a slope; if four-cornered, to come to a standstill, but if round-shaped, to go rolling down.

23. Thus the military force developed by good soldiers is like the momentum of a round stone rolling down a mountain thousands of feet high. So much on the subject of energy.

VI - Weak Points and Strong

1. Sun Tzu said: Whoever is first on the field and awaits the arrival of his enemy will be fresh for the fight; whoever is second on the field and must hasten into battle will arrive exhausted.

2. Therefore, the clever combatant imposes his will on the enemy and does not allow the enemy's will to be imposed on him.

3. By holding out advantages to him, the military leader can cause the enemy to approach of his own accord; or, by creating obstacles, he can make it impossible for the enemy to draw near.

4. If the enemy is taking his ease, he can harass him; if well supplied with food, starve him out; if quietly encamped, he can force him to move.

5. Appear at points where the enemy must hasten to defend; march swiftly to places where he does not expect you.

6. An army may march great distances without distress if it marches through country where no enemy is about.

7. You can ensure the success of your attacks if you only attack undefended places. You can ensure the safety of your defense if you only hold positions that cannot be attacked.

8. Hence a general is skillful in attack whose opponent does not know what to defend; and he is skillful in defense whose opponent does not know what to attack.

9. So the art of subtlety and secrecy! Through it we learn to be invisible, through it inaudible; and hence we hold the enemy's fate in our hands.

10. You may advance and be absolutely irresistible if you attack the enemy's weak points; you may retire safe from all pursuit if you move more rapidly than the enemy.

11. If we wish to fight, we can force the enemy into an engagement even though he is sheltered securely behind a high rampart and a deep ditch. All we need do is attack some other place that will he will be obliged to relieve.

12. If we do not wish to fight, we can prevent the enemy from engaging us even though the lines of our encampment have no fortified defenses. All we need to do is to throw something odd and unaccountable in his way.

13. By discovering the enemy's dispositions and remaining invisible ourselves, we can keep our forces concentrated, while the enemy's must be divided.

14. We can form a single united body, while the enemy must split up into many parts. Hence we will have our whole pitted against his separate parts of a whole, which means that we shall be many to the enemy's few.

15. And if we are able thus to attack his inferior force with our superior one, our opponent will be in dire straits.

16. The place where we intend to fight must not be made known; for then the enemy will have to prepare against a possible attack at several different points; and his forces being thus dispersed in many directions, the numbers we shall face at any given point will be proportionately few.

17. For should the enemy strengthen his front, he will weaken his rear; should he strengthen his rear, he will weaken his front; should he strengthen his left, he will weaken his right; should he strengthen his right, he will weaken his left. If he sends reinforcements everywhere, he will be weak everywhere.

18. Numerical weakness comes from having to prepare against many possible attacks. Numerical strength comes from compelling our enemy to make these preparations against us.

19. Knowing the place and the time of a coming battle, we may concentrate our forces from the greatest distances for the fight.

20. But if neither time nor place be known, then the left wing will be unable to assist the right, the right equally unable to assist the left, the front unable to relieve the rear, or the rear to support the front. How much more so if the furthest portions of the army are anything close to a hundred *li* apart, and even the nearest are separated by several *li*!

21. Though according to my estimate, the soldiers of Yueh exceeded our own in number, that number did not give them an advantage in the matter of victory. I say then that victory can be achieved.

22. Though the enemy has greater numbers, we may prevent him from fighting. Scheme so as to discover his plans and the likelihood of their success.

23. Rouse him to learn the usual methods of his response. Force him to reveal himself, so as to find out his vulnerable spots.

24. Carefully compare the opposing army with your own so that you may know where either army has strength and where either army is deficient.

25. To achieve the best tactical dispositions, conceal them; conceal your dispositions, and you will be safe from the prying eyes of the subtlest spies, from the intrigues of the cleverest minds.

26. Now victory may be produced for your army from the tactics used by the enemy, though they as the multitude cannot comprehend how victory was actually achieved.

27. All men can see the tactics whereby I conquered my enemy, but none can see the strategy I used to create the victory.

28. Do not repeat the tactics that have gained you one victory, but let your methods be devised from the infinite variety of circumstances.

29. Military tactics are like unto water; for water in its natural course runs away from high places and hastens downwards.

30. So in war, the way to fight is to avoid what is strong and attack what is weak.

31. Water shapes its course according to the nature of the ground over which it flows; the soldier works out his victory in relation to the enemy he is facing.

32. Therefore, just as water retains no constant shape, so in warfare, there are no constant conditions.

33. He who can modify his tactics in relation to his opponent, and thereby succeed in winning, may be called a heaven-born captain.

34. The five elements are not always equally predominant; the four seasons make way for each other in turn. There are short days and long; the moon has its periods of waning and waxing.

VII - Maneuvering

1. Sun Tzu said: In war, the general receives his orders from the sovereign.

2. Having then collected an army and concentrated his forces, he must organize the different elements into their units before setting up a proper encampment.

3. After that comes the most difficult task of maneuvering for tactical advantage. The difficulty of maneuvering for tactical advantage consists of turning indirect actions into the direct actions, and misfortune into gain.

4. Thus, to take an indirect route, after enticing the enemy out of the way, and though starting out after him, to contrive to reach the goal before him, you show knowledge of the stratagem of *deviation*.

5. Maneuvering with a disciplined army is advantageous; maneuvering with an undisciplined multitude, most dangerous.

6. If you send a fully equipped army on the march in order to pursue an advantage, the chances are high that you will arrive too late. On the other hand, to detach a small task force for the purpose of pursuing the advantage involves the sacrifice of its baggage and stores.

7. Thus, if you order your men to roll up their buff-coats, and make forced marches without halting day or night, covering double the usual distance at a stretch, doing a hundred *li* in order to wrest an advantage, the leaders of all your three divisions will fall into the hands of the enemy.

8. The stronger men will be in front, the tired ones will fall behind, and on this plan, only one-tenth of your army will reach its destination.

9. If you march fifty *li* in order to outmaneuver the enemy, you will lose the leader of your first division, and only half your force will reach the goal.

10. If you march thirty *li* in order to outmaneuver the enemy, two-thirds of your army will reach the goal.

11. We may take it then that an army without its baggage-train is lost; without provisions it is lost; without bases of supply it is lost.

12. We cannot enter into alliances until we know the intentions of our neighbors.

13. We are not fit to lead an army on the march unless we are familiar with the face of the country—its mountains and forests, its pitfalls and precipices, its marshes and swamps.

14. We cannot turn a natural advantage into a military advantage unless we make use of local guides.

15. In war, practice the art of deception and you will succeed. Move only if there is a real advantage to be gained.

16. Concentrate or divide your troops in accord with the circumstances.

17. Let your rapidity be that of the wind, your appearance imposing as a great forest.

18. In raiding and plundering be like fire; in immovability be like a mountain.

19. Let your plans be dark and impenetrable as night, and when you move, fall upon your enemy like a thunderbolt.

20. When you plunder a countryside, let the spoil be divided amongst your men; when you capture new territory, cut it up into allotments for the benefit of the soldiery.

21. Ponder and deliberate before you make a move.

22. He will conquer who has learnt the art of deviation. Such is the art of maneuvering.

23. The Book of Army Management says: On the field of battle, the spoken word does not carry far enough: hence the institution of gongs and drums. Nor can soldiers clearly see ordinary objects: hence the institution of banners and flags.

24. Gongs and drums, banners and flags, focus the eyes and ears of the army on particular points.

25. The army thus forms a single united body, where it is impossible for either the brave to advance alone, or the cowardly to retreat alone. This is the art of handling large masses of men.

26. When fighting by night, use mostly signal-fires and drums, and when fighting by day, use mostly flags and banners, as a means of signaling your army through sight and sound.

27. Now an entire army may be robbed of its spirit; and a commander-in-chief may be robbed of his presence of mind.

28. A soldier's spirit is keenest in the morning; by noonday it begins to flag; and in the evening, his mind is bent only on returning to camp.

29. A clever general, therefore, avoids an army when its spirit is keen, but attacks it when it is sluggish and inclined to return to camp. This is the art of studying moods.

30. To maintain discipline and calm in an army and await the appearance of disorder and hubbub amongst the enemy: —this is the art of retaining self-control.

31. To be near the goal while the enemy is still far from it, to wait at ease while the enemy is toiling and struggling, to be well-fed while the enemy is famished: —this is the art of conserving one's strength.

32. To refrain from intercepting an enemy whose banners are in perfect order, to refrain from attacking an army drawn up in a calm and confident array: — this is the art of studying circumstances.

33. It is a military principle not to advance uphill against the enemy, nor to oppose him when he comes downhill.

34. Do not pursue an enemy who simulates flight; do not attack soldiers whose temper is keen.

35. Do not swallow bait offered by the enemy. Do not interfere with an army that is returning home.

36. When you surround an army, leave an outlet free. Do not press a desperate foe too hard.

37. Such is the art of warfare.

VIII - Variation in Tactics

1. Sun Tzu said: In war, the general receives his orders from the sovereign, collects his army and concentrates his forces.

2. When in difficult country, do not encamp. In country where high roads intersect, join forces with your allies. Do not linger in dangerously isolated positions. In hemmed-in situations, resort to stratagem. In a desperate position, fight.

3. There are roads that must not be followed, armies that must not be attacked, cities that must not be besieged, positions that must not be contested, orders from the sovereign that must not be obeyed.

4. The general who thoroughly understands the advantages of tactical variations knows how to handle his troops in war.

5. The general who does not understand tactical variations, though he be well acquainted with the face of the countryside, yet will he not be able to turn his knowledge into a practical advantage.

6. So, the practitioner of war who is not adequately trained in the art of varying his plans, even though he knows about advantages, will fail to make the best use of his men.

7. Hence, the wise leader will consider both advantages and disadvantages in his plan.

8. If in our expectations of advantage we also prepare for disadvantages, we may succeed in accomplishing the essential part of our plans.

9. And if, in the midst of difficulties, we are always ready to seize an advantage, we may extricate ourselves from an unfavorable situation.

10. Reduce the hostile chiefs by inflicting damage on them. Make trouble for them, and keep them constantly engaged. Hold out false gains, and make them rush to any given point in their pursuit.

11. The art of war teaches us to not rely on the chance that the enemy will not come, but on our own preparations to receive him; not on the chance of an enemy not attacking, but rather on the fact that we have made our position unassailable.

12. There are five dangerous faults that may affect a general:

(1)   Recklessness, which leads to destruction
(2)   Cowardice, which leads to capture
(3)   A hasty temper, which can be provoked by insults
(4)   A delicacy of honor which is sensitive to shame
(5)   Too much compassion for his men, which exposes him to worry and trouble

13. These are the five dangerous faults of a general, ruinous to the conduct of war.

14. When an army is overthrown and its leader slain, the cause will surely be found among these five dangerous faults. Let them be a subject of meditation.

IX - The Army on the March

1. Sun Tzu said: We come now to the question of encamping the army, and observing signs of the enemy.   Pass quickly over mountains, and keep close to the valleys.

2. Camp in high places, facing the sun. Do not climb heights in order to fight. So much for mountain warfare.

3. After crossing a river, get far away from it.

4. When an invading force crosses a river in its onward march, do not advance to meet it in mid-stream. Let half the army get across, and then deliver your attack.

5. If you are anxious to fight, do not meet the invader near a river he must cross.

6. Take a position on the higher ground facing the sun. Do not move up-stream to meet the enemy. So much for river warfare.

7. When crossing salt marshes, your sole concern is to move through them quickly and avoid any delay.

8. If forced to fight in a salt marsh, you should have water and grass near you and your backs to a stand of trees. So much for operations in salt-marshes.

9. In dry, level country, take an easily accessible position with rising ground to your right and on your rear, so that the danger you face will come from the front and safety will lie behind. So much for campaigning in flat country.

10. These are the four useful branches of military knowledge that enabled the Yellow Emperor to vanquish four rival sovereigns.

11. All armies prefer high ground to low ground and sunny places to dark places.

12. If you care for your men's welfare and camp on hard ground, the army will stay free from disease of every kind. This will spell victory.

13. When you come to a hill or a bank, occupy the sunny side, with the slope on your right rear. Thus you will at once act for the benefit of your soldiers and use the natural advantages of the ground.

14. When heavy rains up-country cause a river that you wish to ford to swell and foam, wait until the waters subside.

15. Country where there are precipitous cliffs, steep ravines, deep natural hollows, confined places, tangled thickets, quagmires and crevasses, should be left with all possible speed and not approached.

16. While we keep away from such places, we should entice the enemy to approach them; while we face them, we should let the enemy have them on his rear.

17. If, in the neighborhood of your camp, there should be any hilly country, ponds surrounded by aquatic grass, hollow basins filled with reeds, or woods with thick undergrowth, they must be carefully and continuously searched; for these are places where men set ambushes and where dangerous spies lurk.

18. When the enemy is nearby yet remains quiet, he is relying on the natural strength of his position for advantage.

19. When he acts aloof and tries to provoke a battle, he is anxious for his enemy to advance.

20. If his place of encampment is easy to access, he is offering a bait.

21. Movement amongst the trees of a forest shows that the enemy is advancing. The appearance of a number of screens in the midst of thick grass means that the enemy wants to make us suspicious.

22. The rising of birds in their flight is the sign of an ambush. Startled beasts indicate that a sudden attack is coming.

23. When there is dust rising in a high column, it is the sign of chariots advancing; when the dust is low, but spread over a wide area, it foreshadows the approach of infantry. When dust branches out in different directions, it shows that parties have been sent to collect firewood. A few clouds of dust moving to and fro signify that the army is encamping.

24. Humble words and increased preparations are signs that the enemy is about to advance. Violent language and driving forward as if to the attack are signs that he will retreat.

25. When light chariots appear first and take up a position on the wings, it is a sign that the enemy is forming for battle.

26. Peace proposals unaccompanied by a sworn treaty indicate a plot.

27. When there is much running about and the soldiers fall into rank, it means that the critical moment has arrived.

28. When some are seen advancing and some retreating, it is a lure.

29. When the soldiers stand leaning on their spears, they are faint from want of food.

30. If those sent to draw water begin by drinking themselves, the army is suffering from thirst.

31. If the enemy sees an advantage to be gained and makes no effort to secure it, the soldiers are exhausted.

32. If birds gather on any spot, it is unoccupied. Clamor by night indicates nervousness.

33. If there is disturbance in the camp, the general's authority is weak. If the banners and flags are shifted about, sedition is afoot. If the officers are angry, it means the men are weary.

34. When an army feeds its horses with grain and kills its cattle for food, and when the men do not hang their cooking-pots over the campfires, showing

they will not return to their tents, you may know that they are determined to fight to the death.

35. The sight of men whispering together in small groups or speaking in subdued tones points to disaffection amongst the rank and file.

36. Too frequent rewards signify that the enemy is at the end of his resources; too many punishments betray a condition of dire distress.

37. To begin by haranguing, but afterwards to be fearful of his men shows a supreme lack of a commander's ability.

38. When envoys are sent with compliments in their mouths, it is a sign that the enemy wishes for a truce.

39. If the enemy's troops march up angrily and remain facing ours for a long time without either joining battle or taking themselves off again, the situation is one that demands great vigilance and circumspection.

40. If our troops are no more in number than the enemy, that is amply sufficient; it only means that he cannot make a direct attack. We simply concentrate all our available strength, keep a close watch on the enemy, and obtain reinforcements.

41. He who exercises no forethought but makes light of his opponents is sure to be captured by them.

42. If soldiers are punished before they are loyal to you, they will not prove submissive; and, unless submissive, they will prove practically useless. If, when the soldiers are loyal to you, punishments are not enforced, they will still be useless.

43. Therefore treat soldiers in the first instance with humanity, but keep them under control by means of iron discipline. This is a certain road to victory.

44. If, when training soldiers, commands are consistently enforced, the army will be well disciplined; if not, the army will have poor discipline.

45. If a general shows confidence in his men but always insists on his orders being obeyed, the gain will be mutual.

X - Terrain

1. Sun Tzu said: We may distinguish six kinds of terrain:

(1)   Accessible ground
(2)   Entangling ground
(3)   Temporizing ground
(4)   Narrow passes
(5)   Precipitous heights
(6)   Positions at a great distance from the enemy

2. Ground that can be freely traversed by both sides is called *accessible* ground.

3. When on *accessible* ground, occupy the raised and sunny areas, and carefully guard your supply lines. Then you will be able to fight with advantage.

4. Ground that can be abandoned but is hard to re-occupy is called *entangling* ground.

5. From *entangling* ground, if the enemy is unprepared, you may sally forth and defeat him. But if the enemy is prepared for your attack, and you fail to defeat him, then disaster will result because you cannot return.

6. When the position is such that neither side will gain by making the first move, it is called *temporizing* ground.

7. When on *temporizing* ground, even though the enemy offers us attractive bait, we do not advance, but rather retreat, thus enticing the enemy in his turn to advance. Then, when part of his army advances, we deliver our attack with advantage.

8. With regard to *narrow passes*, if you can occupy them first, garrison them strongly and await the enemy's arrival.

9. Should the enemy keep you from occupying a pass, do not go attack him if the pass is fully garrisoned. Attack him only if it is weakly garrisoned.

10. With regard to *precipitous heights*, if you are there before your adversary, you should occupy the raised and sunny spots, and there wait for him to come up.

11. If the enemy has occupied *precipitous heights* before you, do not follow him, but retreat and try to entice him back down.

12. If you are situated at a great distance from the enemy, and the strength of your two armies is equal, it is not easy to provoke a battle, and fighting will be to your disadvantage.

13. These six are the principles connected with Earth. The general who has attained a responsible post must study them carefully.

14. Now an army is exposed to six calamities, outside of those arising from natural causes, which are the faults of the generals responsible for them. These are:

(1)   Flight
(2)   Insubordination
(3)   Collapse
(4)   Ruin
(5)   Disorganization
(6)   Rout

15. Other conditions being equal, if one force is hurled against another ten times its size, the result will be the *flight* of the former.

16. When the common soldiers prove too strong and their officers prove too weak, the result is *insubordination*. When the officers prove too strong and the common soldiers prove too weak, the result is *collapse*.

17. When the higher officers are angry and insubordinate, and give battle to the enemy on their own account from a feeling of resentment before the commander-in-chief can tell whether or not he is in a position to fight, the result is *ruin*.

18. When the general is weak and without authority; when his orders are not clear and distinct; when there are no fixed duties assigned to officers and men, and the ranks are formed in a slovenly haphazard manner, the result is utter *disorganization*.

19. When a general, unable to estimate the enemy's strength, allows an inferior force to engage a larger one, or hurls a weak detachment against a powerful one, and neglects to place picked soldiers in the front rank, the result must be *rout*.

20. These are six ways of courting defeat, which must be carefully noted by the general who has attained a responsible post.

21. The natural formation of the country is the soldier's best ally; but a power of estimating the adversary, of controlling the forces of victory, and of shrewdly calculating difficulties, dangers and distances, constitutes the test of a master general.

22. He who knows these things, and in fighting puts his knowledge into practice, will win his battles. He who does not know these things, nor practices them, will surely meet defeat.

23. If a fight is sure to result in victory, then you must fight, even though the ruler forbids it; if a fight will not result in victory, then you must not fight, even at the ruler's bidding.

24. The general who advances without seeking fame and retreats without fearing disgrace, whose only thought is to protect his country and do good service for his sovereign, is the jewel of the kingdom.

25. Regard your soldiers as your children, and they will follow you into the deepest valleys; look upon them as your own beloved sons, and they will stand by you in the face of death.

26. If, however, you indulge your men but are unable to make your authority felt; if you prove kind-hearted but unable to enforce your commands; and if you prove incapable, moreover, of quelling disorder: then your soldiers will act like spoilt children, useless for any practical purpose.

27. If we know that our own men are in a condition to attack, but are unaware that the enemy is not open to attack, we have gone only halfway towards victory.

28. If we know that the enemy is open to attack, but are unaware that our own men are not in a condition to attack, we have gone only halfway towards victory.

29. If we know that the enemy is open to attack, and also know that our men are in a condition to attack, but we are unaware that the nature of the ground makes fighting impractical, then we have still gone only halfway towards victory.

30. Hence the experienced soldier, once in motion, is never bewildered; once he has broken camp, he is never at a loss.

31. Hence the saying: If you know the enemy and know yourself, your victory will not stand in doubt; if you know Heaven and know Earth, you may make your victory complete.

XI - The Nine Situations

1. Sun Tzu said: The art of war recognizes nine types of ground:

(1)   Dispersive ground
(2)   Facile ground
(3)   Contentious ground
(4)   Open ground
(5)   Ground of intersecting highways
(6)   Serious ground
(7)   Difficult ground
(8)   Hemmed-in ground
(9)   Desperate ground

2. When a chieftain fights in his own territory, he is on dispersive ground.

3. When he has penetrated into hostile territory, but to no great distance, he is on facile ground.

4. Ground that will give the side that possesses it a great advantage is contentious ground.

5. Ground on which each side has the freedom to move about is open ground.

6. Ground that forms a key junction to three contiguous states, so that he who occupies it first commands most of the Empire, is ground of intersecting highways.

7. When an army has penetrated into the heart of a hostile country, and has left a number of fortified cities in its rear, it is on serious ground.

8. Mountain-forests, rugged slops, marshes and quagmires—all country that is hard to traverse, is difficult ground.

9. Ground that is reached through narrow gorges, and from which we can retire only through difficult and treacherous paths, ground that is narrow so that a small number of the enemy would suffice to defeat a large body of our men, is hemmed in ground.

10. Ground on which only immediate fighting can save us from destruction is desperate ground.

11. Avoid fighting on dispersive ground. On facile ground, do not halt. On contentious ground, do not attack.

12. On open ground, do not try to block the enemy's way. On the ground of intersecting highways, form alliances with your neighbors.

13. On serious ground, gather your forces and plunder.

14. On hemmed in ground, resort to stratagem. On desperate ground, fight.

15. Those who were called skillful leaders of old knew how to drive a wedge between the enemy's front and rear, knew how to prevent co-operation between his large and small divisions, and knew how to hinder good troops from rescuing bad troops or the officers from rallying their men.

16. When the enemy's men were scattered, they prevented them from concentrating. Even when their forces were united, they managed to keep them in disorder.

17. When it proved to their advantage, they advanced; when otherwise, they stopped still.

18. If asked how to cope with a great number of the enemy arrayed in orderly fashion and in the act of marching to the attack, I should say: "Begin by seizing something which your opponent holds dear; then he will be amenable to your will."

19. Speed is the essence of war: take advantage of the enemy's unpreparedness, make your way by unexpected routes, and attack unguarded places.

20. The following are the principles an invading force should observe: The further you penetrate into a country, the greater will be the solidarity of your troops, and thus the defenders will not prevail against you.

21. Make forays into the fertile countryside in order to supply your army with food.

22. Carefully study the well-being of your men, and do not overtax them. Concentrate your energy and conserve your strength. Keep your army on the move, and devise unfathomable plans.

23. Throw your soldiers into positions where the have no escape, and they will prefer death to flight. If they will face death, there is nothing they may not achieve. Officers and men alike will put forth their uttermost strength.

24. Soldiers in desperate straits lose their sense of fear. If there is no place for them to take refuge, they will stand firm. If they are in hostile country, they will show a tenacious front. If there is no help expected for them, they will fight hard.

25. Thus, without waiting to be marshaled, the soldiers will be constantly on the alert; without waiting to be asked, they will do your will; without restrictions, they will be faithful; without giving orders, they can be trusted to act to your intent.

26. Prohibit the taking of omens, and do away with superstitious doubts. Then, until death itself comes, no calamity need be feared.

27. If our soldiers are not overburdened with money, it is not because they dislike riches; if their lives are not unduly long, it is not because they prefer short lives.

28. On the day they are ordered out to battle, your soldiers may weep, those sitting up crying into their garments, and those lying down letting the tears run down their cheeks. But let them once be in a situation where they cannot take refuge, and they will display the courage of a Chu or a Kuei.

29. The skillful tactician may be likened to the *shuai-jan*. Now the *shuai-jan* is a snake that is found in the Ch'ang mountains. Strike at its head, and you will be attacked by its tail; strike at its tail, and you will be attacked by its head; strike at its middle, and you will be attacked by both head and tail.

30. Asked if an army can be made to imitate the *shuai-jan*, I should answer, Yes. For the men of Wu and the men of Yueh are enemies; yet if they are crossing a river in the same boat and are caught by a storm, they will come to each other's assistance just as the left hand helps the right.

31. Hence it is not enough to put one's trust in the tethering of horses and the burying of chariot wheels in the ground.

32. The principle on which to manage an army is to set up one standard of courage which all must reach.

33. How to make the best of both strong and weak is a question involving the proper use of ground.

34. Thus the skillful general conducts his army just as though he were leading a single man by the hand.

35. It is the business of a general to be discrete to ensure secrecy, upright and just, and thus to maintain order.

36. He must be able to mystify his officers and men by false reports and appearances, and thus keep them in ignorance of his greater plans.

37. By altering his arrangements and changing his plans, he keeps the enemy without definitive knowledge of his intent. By shifting his camp and taking circuitous routes, he prevents the enemy from anticipating his purpose.

38. At the critical moment, the leader of an army acts like one who has climbed up a height and then kicks away the ladder behind him. He carries his men deep into hostile territory before he shows his hand.

39. He burns his boats and breaks his cooking-pots; like a shepherd driving a flock of sheep, he drives his men this way and that, and no one knows where he is taking them.

40. To muster his army and bring it into danger may be termed the business of the general.

41. The different actions suited to the nine varieties of ground, the expediency of aggressive or defensive tactics, and the fundamental laws of human nature, are things that must be studied.

42. When invading hostile territory, as a general principle, to penetrate deeply brings cohesion to the army; to penetrate but a short way means dispersion.

43. When you leave your own country behind, and take your army across neighboring territory, you find yourself on critical ground. When there are means of communication on all four sides, the ground is one of intersecting highways.

44. When you penetrate deeply into a country, it is serious ground. When you penetrate but a little way, it is facile ground.

45. When you have the enemy's strongholds on your rear, and narrow passes in front, it is hemmed-in ground. When there is no place of refuge at all, it is desperate ground.

46. Therefore, on dispersive ground, I would inspire my men with unity of purpose. On facile ground, I would see that there are close connections between all parts of my army.

47. On contentious ground, I would hurry up my rear.

48. On open ground, I would keep a vigilant eye on my defenses. On ground of intersecting highways, I would consolidate my alliances.

49. On serious ground, I would ensure a continuous stream of supplies. On difficult ground, I would push through.

50. On hemmed-in ground, I would block any way of retreat. On desperate ground, I would proclaim to my soldiers the hopelessness of saving their lives.

51. For it is the soldier's disposition to offer a stubborn resistance when surrounded, to fight hard when he cannot help himself, and to obey promptly when he has fallen into danger.

52. We cannot enter into alliances with neighboring princes until we are acquainted with their designs. We are not fit to lead an army on the march unless we are familiar with the face of the country—its mountains and forests, its pitfalls and precipices, its marshes and swamps. We shall be unable to turn natural advantages to account unless we make use of local guides.

53. To be ignorant of any of these principles does not befit a warlike prince.

54. When a warlike prince attacks a powerful state, his generalship shows itself by preventing the enemy from concentrating his forces. He overawes his opponents, and prevents their allies from joining against him.

55. Hence he does not strive to ally himself with all who might wish an alliance, nor does he foster the power of any other state. He carries out his own secret designs, keeping his antagonists in awe. Thus he is able to capture their cities and overthrow their kingdoms.

56. Bestow rewards without regard to conventional rules, issue orders without regard to precedent arrangements, and you will be able to handle a whole army as you would handle a single man.

57. Confront your soldiers with a task itself; never let them know your design. When the outlook is bright, bring it before their eyes; but tell them nothing when the situation is gloomy.

58. Place your army in deadly peril, and it will survive; plunge it into desperate straits, and it will come off in safety.

59. For it is precisely when a force has fallen into harm's way that it is in a position to strike a blow for victory.

60. Success in warfare is gained by carefully accommodating ourselves to the enemy's purpose.

61. By persistently hanging on the enemy's flank, we shall succeed, in the long run, in killing the commander-in-chief.

62. This is called the ability to accomplish a thing by sheer cunning.

63. On the day that you take up your command, block the frontier passes, destroy the official tallies, and stop the passage of all emissaries.

64. Be stern in the council-chamber, so that you may control the situation.

65. If the enemy leaves a door open to you, rush in.

66. Forestall your opponent by seizing what he holds dear, and subtly contrive to time his arrival on the ground.

67. Walk in the path defined by rule, and accommodate yourself to the enemy until you can fight a decisive battle.

66. At first, then, exhibit the coyness of a maiden, until the enemy gives you an opening; afterwards emulate the rapidity of a running hare, and your enemy will find it too late to oppose you.

XII - The Attack by Fire

1. Sun Tzu said: There are five ways of attacking with fire. The first is to burn soldiers in their camp; the second is to burn stores; the third is to burn baggage trains; the fourth is to burn arsenals and magazines; the fifth is to hurl dropping fire amongst the enemy.

2. In order to carry out a fire attack, we must have the means available. The material for raising fire should always be kept in readiness.

3. There is a proper season for attacking with fire, and special days for starting a conflagration.

4. The proper season is when the weather is very dry; the special days are those when the moon is in the constellations of the Sieve, the Wall, the Wing or the Cross-bar; for these four are all days of rising wind.

5. When attacking with fire, be prepared to meet five possible developments:

6. (1) When fire breaks out inside an enemy's camp, respond at once with an attack from outside.

7. (2) If there is an outbreak of fire, but the enemy's soldiers remain quiet, bide your time and do not attack.

8. (3) When the force of the flames has reached its height, follow it up with an attack, if that is practicable; if not, stay where you are.

9. (4) If it is possible to make an assault with fire from the outside, do not wait for it to break out from within, but deliver your attack at a favorable moment.

10. (5) When you start a fire, be to windward of it. Do not attack from the leeward.

11. A wind that rises in the daytime lasts long, but a night breeze soon falls.

12. In every army, the five developments connected with fire must be known, the movements of the stars calculated, and a watch kept for the proper days.

13. Hence those who use fire as an aid to the attack show intelligence; those who use water as an aid to the attack gain an accession of strength.

14. By means of water, an enemy may be intercepted, but not robbed of all his belongings.

15. Unhappy is the fate of the commander who strives to win his battles and succeeds in his attacks without exploiting their success and purpose; for the result is a waste of time and general stagnation.

16. Hence the saying: The enlightened ruler lays his plans well ahead; the good general cultivates his resources.

17. Do not move unless you see an advantage; do not use your troops unless there is something to be gained; do not fight unless the contested position is critical.

18. No ruler should put troops into the field because he is enraged; no general should fight a battle simply out of anger.

19. Advance only if it is to your advantage to advance; if not, stay where you are.

20. In time, anger may change to gladness; vexation may change to contentment.

21. But a kingdom that has once been destroyed can never come again into being; nor can the dead ever be brought back to life.

22. Hence the enlightened ruler is heedful, and the good general full of caution. This is the way to keep a country at peace and an army intact.

XIII - The Use of Spies

1. Sun Tzu said: Raising an army of a hundred thousand men and marching them great distances entails heavy loss on the people and a drain on the State's resources. The daily expenditure will amount to a thousand ounces of silver. There will be commotion at home and abroad, and men will drop down exhausted on the highways. As many as seven hundred thousand families will have their labors impeded.

2. Hostile armies may face each other for years, striving for the victory that is decided in a single day. This being so, to remain in ignorance of the enemy's condition simply because one begrudges the outlay of a hundred ounces of silver in honors and compensation is the height of inhumanity.

3. One who acts thus is no leader of men, no present help to his sovereign, no master of victory.

4. Thus, what enables the wise sovereign and the good general to strike and conquer and achieve things beyond the reach of ordinary men is *foreknowledge*.

5. Now this foreknowledge cannot be elicited from spirits; it cannot be obtained inductively from experience, nor can it be obtained by any deductive calculation.

6. Knowledge of the enemy's dispositions can only be obtained from other men.

7. Hence the use of spies, of whom there are five classes:

(1)   Local spies
(2)   Inward spies
(3)   Converted spies
(4)   Doomed spies
(5)   Surviving spies

8. When these five kinds of spies are all at work, none can discover the secret system under which they operate. This is called "divine manipulation of the threads." It is the sovereign's most precious faculty.

9. Having *local spies* means employing the services of the inhabitants of a district.

10. Having *inward spies* means making use of enemy officials as spies.

11. Having *converted spies* means getting hold of the enemy's spies and using them for our own purposes.

12. Having *doomed spies* means doing certain things openly for purposes of deception, and allowing spies to know of them so they report them to the enemy.

13. *Surviving spies*, finally, are those who bring back news about the enemy's camp.

14. Hence there are no more intimate relations in the whole army than those maintained with spies. No relationship should be more liberally rewarded. In no other business should greater secrecy be preserved.

15. Spies cannot be employed usefully without a certain intuitive wisdom.

16. They cannot be managed properly without benevolence and straightforwardness.

17. Without subtle ingenuity of mind, one cannot make certain of the truth of their reports.

18. Be subtle! be subtle! and use your spies for every kind of business.

19. If a spy divulges a secret piece of news before the time is ripe, he must be put to death together with the man to whom he told the secret.

20. Whether the object be to crush an army, to storm a city, or to assassinate an individual, it is always necessary to begin by finding out the names of the attendants, the aides-de-camp, the doorkeepers, and the sentries of the general in command. Our spies must be commissioned to ascertain these.

21. The enemy's spies, who have come to spy on us, must be sought out, tempted with bribes, led away, and comfortably housed. Thus they will become converted spies and be available for our service.

22. It is through the information brought by the converted spy that we are able to acquire and employ local and inward spies.

23. It is owing to the information of converted spies, again, that we can cause the doomed spy to carry false tidings to the enemy.

24. Lastly, it is by this information that the surviving spy can be used on appointed occasions.

25. The end and aim of spying in all its five varieties is knowledge of the enemy; and this knowledge can only be derived, in the first instance, from the converted spy. Hence it is essential that the converted spy be treated with the utmost liberality.

26. Of old, the rise of the Yin dynasty was due to I Chuh who had served under Hsia. Likewise, the rise of the Chou dynasty was due to Lu Ya who had served under the Yin.

27. Hence it is only the enlightened ruler and the wise general who will use the highest intelligence of the army for purposes of spying and thereby they achieve great results. Spies are a most important element in water, because on them depends an army's ability to move.

About the Author

Robert Cantrell first studied *Sun Tzu on the Art of War* for its intended purpose while training to be a military officer at Duke University in 1985. He graduated from Duke University in 1987 with degrees in biology and military science and served on active duty with the 101st Airborne Division of the U.S. Army as an infantry platoon leader.

After military service, Robert entered business as a sales representative for IBM/ROLM, Dean Witter, and Thomson Information. During that time, Robert earned his MBA at the Edinburgh Business School of Heriot-Watt University in the United Kingdom. In 1994, Robert began to speak and consult on competitive intelligence and intellectual property strategy throughout North America, Asia, and Europe. By 1999, this practice had expanded to include all aspects of business strategy.

As a part of Robert's professional development, Robert continued his study of Sun Tzu's *Art of War* as well as other classic treatises on strategy, to include Carl von Clausewitz's *On War* and Niccolo Machiavelli's *The Prince*. Through these studies, he has documented the core principles of each treatise as it applies to business, the military, and law.

In 2002, Robert started the company Center For Advantage. Center For Advantage provides products and services to help clients create breakthrough strategies on their own. He also runs a consulting service, Strategy Innovators, designed for experienced clients both willing and able to take actions determined necessary to succeed. For more information on these products and services, see www.centerforadvantage.com and www.strategyinnovators.com.

Center For Advantage

Robert Cantrell provides strategy innovation products and services through his company, Center For Advantage. These products and services are designed to help clients create breakthrough strategies on their own. The consulting arm, Strategy Innovators, is a selective consulting service for experienced clients both willing and able to take actions determined necessary to succeed.

Please address inquiries involving Center For Advantage products and services, plus inquiries concerning speaking and seminar engagements, to the address below. For more information on Center For Advantage products and services, please see www.centerforadvantage.com and www.strategyinnovators.com.

Additional copies of *Understanding Sun Tzu on the Art of War* can be purchased directly from the address below as well. Please enclose a check or money order for $14.95 plus $4.00 for shipping and handling. (Virginia residents add $0.67 per book for Virginia sales tax.) Inquire for volume discounts. If you would like this book sent as a gift, please indicate this and provide the intended destination.

Robert L. Cantrell
Center For Advantage
P.O. Box 42049
Arlington, VA 22204
info@centerforadvantage.com

Acknowledgements

The author would like to express his thanks to:

Lionel Giles, M. A. for his early work to bring an accurate translation of Sun Tzu's text to the English speaking world in 1910. An original copy of his published 1910 translation, which accurately portrays the spirit of Sun Tzu's work, became the basis of this book.

James Legge, whose 1891 translation of the *Tao Te Ching* likewise brought the spirit of Lao Tzu's work to the English speaking world and became a foundation for illustrating the relationship of Sun Tzu's work to Taoist philosophy.

Members of the U.S. Army, and in particular, the Duke University ROTC instructors who provided an introduction to Sun Tzu and the 101st Airborne Division (Air Assault).

Instructors from the National Defense University in Washington D.C. for their suggestions on the text.

The Black Tree History Group for both editorial work and a review of historical points made in the text.

Atchity Entertainment/Editorial International for editing and project management support.

Parents Robert Cantrell and Mary Lou Cantrell for editing and content suggestions.

Julia Goswick for her editing and support.